Advancing Global Bioethics

Volume 20

The book series Global Bioethics provides a forum for normative analysis of a vast range of important new issues in bioethics from a truly global perspective and with a cross-cultural approach. The issues covered by the series include among other things sponsorship of research and education, scientific misconduct and research integrity, exploitation of research participants in resource-poor settings, brain drain and migration of healthcare workers, organ trafficking and transplant tourism, indigenous medicine, biodiversity, commodification of human tissue, benefit sharing, bio-industry and food, malnutrition and hunger, human rights, and climate change.

Joaquim Braga
Editor

Ethics of Care

Theoretical and Practical Perspectives

Springer

Editor
Joaquim Braga
Department of Philosophy
Communication and Information
University of Coimbra
Coimbra, Portugal

ISSN 2212-652X ISSN 2212-6538 (electronic)
Advancing Global Bioethics
ISBN 978-3-031-73607-0 ISBN 978-3-031-73608-7 (eBook)
https://doi.org/10.1007/978-3-031-73608-7

This Springer imprint is published by the registered company Springer Nature Switzerland AG
The registered company address is: Gewerbestrasse 11, 6330 Cham, Switzerland

Introduction

From an ethical standpoint, contemporary society is deeply entangled in a complex network of cultural, political, and economic relations, making it challenging to come up with universal solutions. While similar challenges may arise in different parts of the world, society is still influenced by various regional and national moral frameworks that shape individuals' lives and the communities they belong to. Can these localized moral frameworks effectively address global challenges? Addressing this question shouldn't automatically lead to the conclusion that it's impossible to create an ethical framework that integrates the different moral viewpoints in society. If that were the case, we would continue to struggle between universalism and communitarianism, losing sight of the need to address current issues and potentially obscuring their reality. Maintaining awareness of *facts* is just as relevant as maintaining awareness of *values*.

In the era of hyperinformation, it is crucial to be aware of facts. Facts guide our daily actions, not just inform us. An ethics of care is imperative as it highlights formal and informal care practices that fulfil needs and prevent problems. Shedding light on those everyday care practices also demands a critical examination of social ways of life and their impact on the relationships shaped by care.

Current issues such as those related to the market economy, widespread technological development, and global warming imply a relational deficit in human behaviour and the values that regulate it. This has social, cultural, and political repercussions on the organization of care work in both institutional and private spheres, as well as on the recognition of the status of care workers. The attention given to a care receiver by a caregiver does not always have corresponding social and political attention regarding the work and status of the caregiver. The issue of informal care—particularly domestic work carried out by women—brings into play traditional values shaped by discrimination, which urgently need critical consideration.

The conception of an ethics of care has always involved critical positioning, as evidenced by the influential works of Carol Gilligan (1982), Virginia Held (2006), and Joan Tronto (1994). These studies underline the need to challenge the significant social imbalances perpetuated by current moral principles and the resulting prioritization of care work. It is noteworthy that, at the inception of the ethics of care, issues related to gender and the division of labour based on sex are emphasized and serve as key focal points for criticizing modern society. Merely assuming the intentions of moral agents, as is often done in traditional ethics, is insufficient. Uppermost, it is imperative to acknowledge their social circumstances. Those who provide care are not at all times recognized as *moral agents*; even when they are, their status is not entirely adequately valued for the effective continuation of care practices.

The ethics of care extends beyond its application to specific areas of human action. It functions as a form of *metaethics* by allowing a critical examination of the theoretical principles that underlie behaviour in general and moral behaviour in particular. The concept of "vulnerability", central to the ethics of care, emphasizes the necessity of including a universal ontological assumption in philosophical thinking, encompassing the lives of humans, other living beings, and the planet they inhabit. This approach challenges traditional deontological and consequentialist ethics, stressing that *what already is* must have equal philosophical value to *what should be*.

Global issues such as starvation and the environmental crisis compel us to recognize the far-reaching impact of our actions and the underlying values. A broad causal framework is needed. This fact demands taking ethical responsibility beyond individual moral behaviour. When we cannot foresee all the consequences of our actions, especially in the environment, we must not regard our values as the sole guarantors of life's continuity and sustainability. Rather, our values and those of others require acknowledging that actions, agents, and their effects are not always immediately or directly evident, yet they indelibly shape the world.

A growing body of research on care aims to redefine it as a fundamental cornerstone of transnational justice. This view highlights the pivotal role of politics, institutions, and decision-makers. The way care relations can impact power relations represents the core theoretical challenge in the field of the ethics of care. Virginia Held emphatically argues that "a globalization of caring relations would help enable people of different states and cultures to live in peace, to respect each other's rights, to care together for their environments, and to improve the lives of their children" (Held 2006: 168). According to the author, the idea that ethical principles can globally converge is unlikely to be achieved solely through modern rationalism and liberal contractualism. Instead, it will only prevail in our societies when grounded in a "universal experience" of human behaviour, or what she refers to as "the experience of being cared for" (Held 2006: 132). From Held's perspective, globalized ethics of care "does not require cultural similarity, and care across cultural divides may succeed better than mere liberal tolerance" (Held 2006: 132).

A global ethical framework must be grounded in the philosophical principles of the ethics of care. Recognizing and understanding significant care practices worldwide is a crucial starting point for advocating a globalized ethical approach. With the proliferation of new information and communication technologies, we have access to numerous images of distant events, expanding the scope of our concerns. However, our tendency to focus on negative imagery amplifies the number of problems and, consequently, the moral values needed to solve them. That tendency not only leads to inertia and fatalistic attitudes but equally fails to highlight everyday human practices that embody values contrary to those contributing to such problems. Fiona Robinson raises a pertinent question: if the notion of care embodies "a morality of 'closeness' rather than 'distance', how useful could an ethics of care possibly be, then, when applied to the global context?" (Robinson 1997: 125). She answers this question by asserting that our current era of globalization "is characterized by increasing closeness, new relationships, and profound interdependence". In light of this reality, we must recognize that this "era of global interdependence demands an interpersonal morality—one characterized by attentiveness to needs and concerns and that recognizes others as concrete rather than generalizable" (Robinson 1997: 126).

We need to remember that care is more than just the intention of looking after something. It extends beyond showing concern and having corresponding emotional states. It is the actual practice of caring that gives it its true meaning. When action aligns closely with intention and time and space are intertwined, the ethics of care finds its main theoretical foundation. The expression "landscapes of care", coined by Christine Milligan and Janine Wiles (2010), emphasizes the spatial conditions that involve care relations. Both authors argue that despite the meaningful link between distance and proximity in physical relationships, "even physically distant caregivers can be affectively or socially proximal, and that physically distant care relationships can be literally embodied" (Milligan and Wiles 2010: 749).

The distinctiveness of care practices lies in their power to evoke emotions, regardless of their geographic, cultural, social, and political settings. This affective resonance plays a significant role in mitigating the impact of social and cultural differences and heightening awareness of vulnerability, establishing it as a fundamental pillar of a global ethical approach. Thus, the manner practices are spread and reproduced is just as important, and sometimes even more so, than their moral content.

The imperative application of the moral law to one's own will unequivocally shape the ethical core of modern Western thought. Therefore, attempts to reduce the ethics of care to a family and domestic moral framework are fundamentally mistaken. They falsely assume that the relational nature of human beings in their daily activities lacks prescriptive moral power that can be universalized. However, there are regional care practices that, despite being rooted in specific communities, have far-reaching global impacts. The construction of shared moral values must take those practices into account. Before delving into abstract values, it is vital to scrutinize empirical practices that inherently embody values. Such values are not purely moral. They may initially belong to other axiological orders—for instance, an aesthetic order—, yet effectively contribute to a balanced social life.

The ethical gaze should shift from excluding aspects of social life that cannot be universally applied to embracing practices and values that already exist within certain human cultures. These situated axiological frameworks can promote acceptance of otherness and recognition of the vulnerability of beings and the environment. Instead of relying on the modern universalism *logic of good intentions*, one must acknowledge the diversity of social and cultural realities. In order to establish a global ethical approach, it is necessary to foster trust and mutual respect among communities by identifying and reconciling the differences in their ways of life. The key focus here is to develop a procedural framework that seeks to find common ground between the particular and the general.

The concept of care involves focusing on the broad sphere of responsibility, particularly the attention given to others in our daily actions, even when we don't fully understand the reasons for their actions. Essentially, care involves being ethically responsible for life in the world. Geographical and territorial limits are not the real boundaries of care—instead, the true obstacles to care are imposed by life itself. It's not just about specific cultural values—care practices have a global purpose and create connections between different forms of life, transcending social and cultural differences. Despite such asymmetries, the vulnerability of beings requires a broad ethical approach, emphasizing the fundamental connection between our lives and the lives of others.

It is with all these mutual implications in mind that this volume is divided into two main parts—Care in Theory and Care in Practice. Both parts aim to emphasize the understanding of care ethics as a philosophical domain that relies on interdisciplinary knowledge to form theoretical proposals. The various care practices that have become widespread and, in some cases, gradually disappeared, along with the semantic changes of the term "care", always require a philosophical reflection on the concepts of care and caring. This task will be one of the main focuses of the first part of this volume.

In Chap. 1, the focus is on Paul Ricoeur and Emmanuel Levinas' philosophical thought about the doctor-patient interaction. Medical ethics must consider the need for an open dialogue that promotes trust and responsibility. If suffering and vulnerability are seen as the ultimate expressions of otherness, then caring is the most evident form of hospitality. It's not surprising that authors such as Levinas and Virginia Held consider motherhood as the best metaphor for idealizing care-based social relationships. As noted in Chap. 2, "motherhood" implies foremost responsibility for the vulnerability of others and is not solely limited to female behaviour in a biological sense. Motherhood, as a philosophical metaphor, can be applied to the doctor-patient interaction to establish a therapeutic ethical model based on trust and independent of contractual and economic power relations.

The significance of an open dialogical space, as demonstrated in the first two chapters, is brought to the fore in Chap. 3. Vulnerability seems to be related to attention in the sense that caring activities require an ethical and aesthetic expression of thought and emotion. Without this, others' suffering becomes invisible, to the extent that we overlook it and do not prioritize it in our daily lives. In the realm of justice, there is also an increasing need to highlight human vulnerability. As shown in Chap. 4, this is only truly achievable if care is integrated into a broader understanding of all citizens' rights and responsibilities, as well as the constitutional framework that defines them. We should always bear in mind that building relationships with people, whether near or far, permeates both care practices and the laws themselves.

Chapter 5 focuses on the relational nature of care, qualifying care as a "relational virtue" and resolving the theoretical incompatibilities between care ethics and virtue ethics, as well as the alterity and autonomy of moral agents. The concepts of vulnerability and responsibility do not imply the suspension of one's autonomy. Caring and being cared for are linked to these two concepts (vulnerability and responsibility). However, Chap. 6 shows that there are instances—such as, for example, in psychiatric illnesses—where "radical vulnerability" occurs, bringing individuals dangerously close to the "abyss". In this deep vulnerability, both the caregivers and those being cared for experience an almost symmetrical existential condition. According to Levinas' approach, this also means that ethics takes precedence over ontology.

The chapters mentioned above emphasize the importance of considering the concept of care within a relational and intersubjective theoretical framework. This goes beyond the common perception of care as solely a personal concern. Now, the concluding chapter of the first part of this volume emphasizes such a theoretical framework. Beginning with a critical analysis of Martin Heidegger's ontological notion of care (*Sorge*), as expounded in *Sein und Zeit*, Chap. 7 emphasizes our obligation to care for those who have previously cared for us, including our grandparents. The goal is to give full expression to an intergenerational memory and commitment that goes beyond individualistic and egocentric perspectives on human nature. Providing care to elderly individuals initiates a semantic expansion of the notion of care, imbuing it with intuitive dimensions derived from an involved social life that is close to the reality of other beings.

In the second part of *Ethics of Care*, the questions raised prompt us to emphasize the transnational reach of care practices. Unlike conventional ethical theories, the ethics of care is primarily expressed through individual and social practices that foster human mutual relations, rather than through specific moral values intended to govern those practices. Therefore, any proposals for an ethics of care should first and foremost consider the practical realm of relationships among individuals and between individuals and the world.

Chapter 8 expands the concept of vulnerability to consider the impact of material consumption on nature and interpersonal relationships. Sustainable consumption practices, underpinned by an ethics of care, connect individuals' duty to how the material world is created, maintained, and consumed. However, the chapter argues that "ethical consumption" alone cannot effectively promote sustainability. Without an ethics of care that addresses the interactions between humans and non-humans, we are simply duplicating the logic of the capitalist market. Taking care of what is close to us can mean overlooking what is far away. Chapter 9 demonstrates how toxic waste is consigned to certain areas of the planet that are unseen and outside our focus of attention, leading to increased environmental pollution and social inequities. One strategy for mitigating irresponsible consumerism is to design and manufacture consumer goods under a global ethical paradigm based on care.

In care, the concept of responsibility must be decentered from the family sphere and extended to society as a whole. Chapter 10, which uses Mexican public childcare policies as a case study, argues that when the economic circumstances and sociocultural context of care providers are ignored, a purely biological and maternalist view of childcare is repeated—this can lead to increased poverty and violence. Political actors play an important role in regarding care practices as opportunities for erasing injustices and socioeconomic inequities, rather than simply providing services. Chapter 11 examines the Informal Caregiver Statute in Portugal, with the primary goal of highlighting the necessity for care practices related to female work to be the focus of profound socialization. Addressing the sexual division of labour, which persists in childcare-related tasks, entails demanding "collective responsibility" from public policies that free women from the household burden and satisfy their true needs. The goal here is to bring to light what some deep-rooted care practices may conceal, that is, injustice.

Chapter 12 examines the relationships between care and technology on two intertwined theoretical levels: the first seeks to reflect on ways to maintain the relational character of care in the face of increased technological mediation; the second seeks to demonstrate how formal and informal care practices, when permeated by reciprocity, foster a critical view of the use we make of technological artefacts themselves. Instrumentalist and utilitarian views of technology and ethics can be best disproved by considering this dual link between care and technology. The concluding chapter of this volume strongly emphasizes the critical importance of contemplating the impact of new technologies on environmental care practices. Chapter 13 expands on the premise that if technology is the only cultural form that dictates ethical paths, then the rising ecological concerns that we face now will only have technological solutions. Within this reductionist logic, the term "Anthropocene" alludes to a "techo-care" in which nature is used as both an object and a resource, and it is conceived via a solely technological lens. Geoengineering solutions to environmental problems highlight the argument that only technology can remedy problems generated by technology. Ecological ethics must be founded on caring practices based on a responsibility that safeguards otherness, rather than following the route of moral authoritarianism, which is encouraged by the fantasy of technology's limitless power.

All spaces and contexts where care practices take place—such as hospitals, schools, and nursing homes—and all locations where these practices are currently lacking but should be implemented are crucial for acknowledging and addressing the vulnerability of individuals and the world. Therefore, in this volume, the ethical focus is on understanding what is currently being done and what needs to be done, rather than just underling human intentions as traditional psychology-based ethical approaches often tend to do. This shift in focus is fully justified by the need to prioritize the sociocultural nature of human relationships and the conditions that enable them, rather than simply imposing certain disconnected values that disregard both.

The chapters of this volume stress the relational nature of care. Keeping in view the ideas of human vulnerability and social contingency, they show how an ethics of care supports moral, political, aesthetic, and environmental values on a global scale. This convergence of values goes beyond just moral considerations, as ethics of care can help us understand the broader possibilities of human behaviour from a global perspective. Recognizing vulnerability helps us know how it affects the development of our moral beliefs. Vulnerability and morality are not always connected. On the contrary, they can even be exclusionary if the first hinders or completely breaks the relational character of care and, therefore, calls into question the dimension of otherness inherent to the second.

As the organization of this volume depended on the cooperation of several people and institutions, I would like to express my gratitude to all the authors of the chapters for their exemplary handling of the publication process. I would also like to thank the Editorial Director of Springer Humanities, Floor Hosting, as well as the editors of the Advancing Global Bioethics series, Henk ten Have and Bert Gordijn, for their interest and commitment in publishing this volume. Additionally, I am deeply thankful to Professor Doctor Mário Santiago de Carvalho and my colleagues at the Institute for Philosophical Studies of the University of Coimbra for their support and mentorship of this project.

Coimbra, Portugal Joaquim Braga
September 2023

References

Gilligan, Carol. 1982. *In a different voice: Psychological theory and women's development.* Cambridge, MA: Harvard University Press.

Held, Virginia. 2006. *The ethics of care: Personal, political, and global.* Oxford: Oxford University Press.

Milligan, Christine, and Wiles, Janine. 2010. Landscapes of care. *Progress in Human Geography* 34(6): 736–754. https://doi.org/10.1177/0309132510364556

Robinson, Fiona. 1997. Globalizing care: Ethics, feminist theory, and international relations. *Alternatives: Global, Local, Political* 22(1): 113–133. https://doi.org/10.1177/03043754970220

Tronto, Joan. 1994. *Moral boundaries. A political argument for an ethic of care.* New York/London: Routledge.

Contents

Part I
Care in Theory

Chapter 1
The Relevance of Paul Ricoeur's and Emmanuel Levinas's Thoughts for an Ethics of Medical Care

Lazare Benaroyo

Abstract Since the 1970s, the traditional medical ethics model has been an attempt to understand theoretical biomedical ethics and then apply it to clinical situations. Many practitioners and some philosophers considered this misguided from the start. They tried to understand which kinds of ethical responsibilities flow from a clearer understanding of medical care rather than pull one or another ready-made general theory of ethics off the shelf and apply it to the clinical reality. In the context of contemporary debates on the philosophical foundations of medical practice, this paper explores the anthropological, epistemological and ethical dimensions of clinical work. Furthermore, it endeavours to show the extent to which Paul Ricoeur's and Emmanuel Levinas' conceptions of ethical responsibility may help us grasp medical care's ethical core. It is suggested that this approach could shape an appropriate ethical understanding of medical practice grounded in an ethics of care.

Keywords Ethics of care · Medical care · Ethical responsibility · Vulnerability · Practical wisdom · Paul Ricoeur · Emmanuel Levinas

When in good health, the body is perceived as something self-evident. But when it is ill, when it is cracked, existence can no longer be taken for granted—there is a sense of malaise, and the medical quest begins. How do healthcare professionals respond to this call for help? They adopt a scientific approach, seeking to explain the cause of the body's malfunction, 'thing among things' (Merleau-Ponty 1964, p. 180), and thus control the course of the illness. They also adopt a personal

---

Part of this chapter is translated and adapted from "L'éthique et les paradoxes du soin". In *Les valeurs du soin. Enjeux éthiques, économiques et politiques*, eds. Jean-Philippe Pierron, Didier Vinot et Elisa Chelle, 175–186. 2018. Paris: Seli Arslan. Reproduced with authorisation.

L. Benaroyo (✉)
University of Lausanne, Lausanne, Switzerland
e-mail: lazare.benaroyo@unil.ch

J. Braga (ed.), *Ethics of Care*, Advancing Global Bioethics 20,
https://doi.org/10.1007/978-3-031-73608-7_1

attitude, seeking to understand the meaning of the crack in 'being in the world' that has disrupted the hitherto untroubled harmony of the patient's existence.

As a scientific act, medicine is nowadays seen as the application of a broad range of evidence-based knowledge that serves as a basis for practice. This offers means of therapeutic, preventive and diagnostic action. In the clinical field, this approach is backed by an art in the sense of know-how, referred to as *practical wisdom*. In other words, the healthcare professional's ability to make the most suitable possible decision in a singular healthcare situation. An art or a science? Contemporary medical wisdom attempts to combine the two, which sometimes conflict, and often lead to misunderstandings. What light can ethical and philosophical reflection shed on this issue? To what extent are Paul Ricoeur's and Emmanuel Levinas' thoughts relevant to lay the foundations of an ethics of medical care?

1 A Historical and Epistemological Perspective

Until modern times the art and science of medicine focused on care of the patient. However, since the nineteenth century, the increased importance of scientific approaches has broken this link by dissociating art and science.

When we read the Hippocratic texts, it seems that care was above all seen as an art—*ars medica*—in the sense of a *techne*: a skill, an ability, a craft. This medical know-how was understood as *techne iatrike*. Men of art were 'artisans', driven by *philotechnia* (love of technique) and *philanthropia* (love of human beings). These 'artisans' used meticulous observations to monitor the contextual features that marked the patient's specific situation, his *habitus*, to adapt the choice of remedy. Art and science were focused on the same goal: science at the service of art to provide individualised care was the motto of the Hippocratic ethic (Hippocrates 1999, p. 115).

From the perspective of modern science, recovery of health depends on an ethic of technical skill aimed at correcting measurable biological disorders. Since the mid-nineteenth century, the Hippocratic art of medicine has lost its place within an epistemological approach based on the natural science model. This model draws its strength from the fragmentation of knowledge and the disciplinary differentiation that allows specialisation in the field of organ pathology, setting aside in some way the significance of the patient's proper suffering experience (Jewson 1976). As Georges Canguilhem pointed out:

> The [scientific] method has set aside patients, understood as seeking elective attention to their pathological situation ... Like all sciences, medicine has had to go through the stage of temporarily eliminating its specific initial subject matter (Canguilhem 2002, pp. 419–420 and 428).[1]

[1] The translation of the citation is my own.

In 1926, the professor of medicine at Harvard University, Richard Cabot, stated that it was time for medical science to start focusing on a conception of care understood as a 'curing attitude'. He advocated abandoning an approach based on overall care as caregivers 'attention was "too strongly concentrated" on the difficult tasks of diagnosis and treatment, and 'there is not enough attention to go round'" (Reich (1995 [1], p. 333). According to Cabot, medicine should focus on the body; other facets of care belong to other disciplines 'in establishing the professions of medical social work and clinical pastoral care'. The development of medical science would henceforth be progressively structured by the world of expertise.

However, this view was not universally accepted. It triggered a vigorous reaction from clinicians, who deplored this new conception of care and the erasure of the accompanying art of medicine. In 1927, Cabot's colleague Francis Weld Peabody thus wrote:

> One of the essential qualities of the clinician is interest in humanity, for the secret of the care of the patient is in caring for the patient (Peabody 1927, p. 882).

William Osler joined Peabody in promoting a humanist clinical approach capable of paying equal attention to the technical and human dimensions. He stressed the importance of being able to restore the balance in the organism while correcting organic dysfunction so that patients could once more live lives in which they could fulfil their potential and regain their freedom.

In the mid-twentieth-century post-war setting, this trend faded in response to scientific and technical developments from across the Atlantic. However, in the 1980s it was revived in both Europe and the United States (Schwartz and Wiggins 1985, 1988). Clinicians were faced with a paradox in the closing decades of the twentieth century: care should precede the technical act, whereas at present it is the technical act that guides care.

As clinicians saw it, the paradox lay in the twofold injunction that structures medical care in the nascent technological world: attention to organic pathologies must be based on knowledge of the general, yet attention to the organism as a whole must be based on knowledge of the singularity. How are these two kinds of knowledge to be reconciled? How are we to revive a conception of care that enables medicine to perform its task, a conception that combines science and art, taking account of the organism and the human factor as a whole, while paying particular attention to organ and tissue damage? This question had already been raised in Germany in the 1930s when medical anthropology (*Medizinische Anthropologie*) was emerging (ten Have 1995).

This challenge to contemporary medicine seems hard to overcome in a highly technological world. How are we to create the conditions that will allow healthcare professionals to listen to, and at the same time to hear, what patients are saying (Baron 1985)? At the end of the 1980s, this question led to reflections heralding an ethical approach that would serve as the foundation for the development of a philosophy of care (Lefève 2014).

As Michael Schwartz and Osborne Wiggins point out (Schwartz and Wiggins 1985), dealing with these two conceptions is not self-evident, for they are first and foremost paradoxical injunctions that call for different kinds of knowledge and methods. Health care providers' responsibility calls on them to adapt their technical expertise (based on mastery and control) to be receptive, welcoming and listening (which presupposes self-effacement, lack of mastery, and openness) to effectively care for others. Contemporary clinicians are simultaneously confronted with the universal and the singular, the objective and the subjective, and the wish for infinity in the presence of human finiteness. They thus face the need to give their technical practice a place in the care plan, while accepting that the scientific approach encounters its limitations in the clinical setting. Namely, the difficulty it has in understanding other people's suffering, and in grasping the profound issues of otherness and singularity that are associated with it. As Canguilhem wrote:

> We have to admit to ourselves … that there can be no homogeneity and uniformity in attention and attitude towards the disease a patient has and the patient's illness experience, and that caring for a patient does not involve the same responsibility as rationally fighting disease (Canguilhem 2002 [1968], p. 408).[2]

Canguilhem therefore stresses the importance of being attentive to the "point of conversion of the value of medical rationality", a crucial point "that is not a point of withdrawal" (Canguilhem 2002 [1968], p. 411). He believes that, in combination with the scientific approach directed towards the translation of symptoms into objective signs, physicians should be able to 'step aside'. This step will allow them access to the interpretative aspect of a patient's own experience of their illness. Patients interpret their symptoms in their registers of interpretation of the world and themselves, and they call on physicians to explain them based on their own words and narratives. As Céline Lefève argues in a commentary on Canguilhem's work:

> Failing this, a register of medicine (rather than an addition to it) is missed, adding new sufferings to the pathological ones. Ignoring patients' subjective experiences adds emotional harm to the risk of diagnostic and therapeutic error' (Lefève 2014, p. 211).[3]

This 'step aside', this shift in physicians' attention from disease knowledge to illness as vital experience (Lefève 2014, p. 213), is closely linked to their moral commitment as health carers. It echoes the twofold reference to our conception of the body that is pointed out by Maurice Merleau-Ponty when he says that our body is a "two-sided being" that contains "a double belongingness to the order of the 'object' and to the order of the 'subject'" (Merleau-Ponty 1964, pp. 180–181). Mirroring this, Canguilhem stresses that the status of contemporary medicine lies in its *dual character*. Therefore, Medicine must create a link between 'knowledge and understanding inherent in clinical work' (Lefève 2014, p. 213).

How can one gain access to this 'conversion of medical rationality', which is the source of 'the shift of focus in the physician's attention' towards an approach to medical care that pays attention to the individual singular distress and needs? To what extent are Ricoeur's and Levinas' thoughts relevant to this aim?

---

[2] The translation of the citation is my own.

[3] The translation of the citation is my own.

2 An Ethical and Philosophical Approach to Medical Care: Two Ways of Shifting Focus

1. **Paul Ricœur's approach**

Paul Ricœur's work on medical judgment seems particularly relevant when responding to this challenge, for it provides the conditions for overcoming the dual character of contemporary medicine as part of a revival of Aristotelian 'practical wisdom'. Ricœur defines 'practical wisdom' (Ricoeur 2000, pp. 16–26) as the result of deliberation exploring the following three spheres of action: interpersonal, professional and socio-cultural. Each of these spheres refers to a different epistemological, anthropological and ethical register. According to Ricœur, a detailed, individualised prudential judgment, at the basis of practical wisdom, can be constructed by passing through the three stages of action that pervade these three spheres: the stage of the ethical aim of restoring the patient's "being-able-to-be-there", the stage of technical intervention and the stage of personalised care. This enables the technical and existential registers to be linked up in a care plan that pays attention to the patient's singularity.

In Ricœurian terms, the three levels of medical judgment are:

- the level of prudence at which the ethical aim of the care plan is revealed through intimate dialogue with the patient—the teleological stage at the basis of the first-order prudential judgment.
- the deontological level—the technical stage at the basis of the deontological judgment.
- the reflexive level—the point at which the first-order prudential judgment is actualised, the moral judgment in the given situation forming the basis of practical wisdom.

Let us consider how these levels of medical judgment are linked to help elaborate practical wisdom through several 'shifts of focus'.

In his article, Ricœur explains the meaning of the health carer's *ethical aim* in the field of clinical practice. In his view, this is asymmetrical from the very outset; it seeks to restore the patient's "being-able-to-be-there": (a) the complaint, which involves the patient, and (b) the deployment by the health carer of the resources that will help restore the patient's "being-able-to-be-there". According to Ricœur, the patient is confronted with an illness, and the physician has the qualifications to remedy it. This aim has its source in the patient's experience, most often 'based upon a social relation for which suffering is the primary motivation and has as its *telos* the hope of being helped and perhaps cured' (Ricoeur 2000, p. 16). To Ricœur, the ethical core of this encounter is a *pact of care based on trust* sealing an original process whereby the initial asymmetry of the encounter is to be overcome on the existential level.

From this point of view, the health carer's ethical aim commits him to grasp what suffering means for the patient. This approach is based on an understanding of the specific links between suffering and personal identity in the context of illness. It is part of an approach whose focus is shifted away from the classic case history as an expression of medical rationality. Here, the attention to the narrative is crucial.

To Ricœur, however, this conception of the ethical aim can only become relevant if it has passed through what he calls the 'sieve of the norm', guaranteeing that the face-to-face relationship with others or oneself within a technical world takes place under the sign of justice (Ricoeur 1992, pp. 203–239). Enriched by having passed through the sieve of the norm, the ethical aim should guide the moment of practical wisdom. It then, says Ricœur, becomes relevant by focusing on returning to attentiveness to the singularity of situations (Ricoeur 1992, pp. 240–296).

Practical wisdom completes the process of making the medical judgment at the end of a deliberation. In a reflexive movement that extends beyond technical and disciplinary expertise, practical wisdom explores the basic values that underpin conflicting deontological and first-order prudential judgements. According to Ricœur, the reflexive level opens the way to the possibility of coping with the 'tragedy of action'.

Making an actualised prudential judgement then depends on the health carer being willing and able to give a preponderant position to interdisciplinary dialogue—the only way, says Ricœur, to strike a "well-considered balance between conviction and argument" (Ricoeur 1992, p. 289). The actualised prudential judgment, referred to as 'critical solicitude' (Ricoeur 1992, p. 273), should ideally be able to take place within a 'cell of good counsel' based on the 'model of a small circle that brings together relatives, physicians, psychologists and clergy at a patient's bedside'. Wisdom of judgment, says Ricœur, should always be practised, and judgments based on wisdom be made, by several people. Such 'deliberation by several people', the second way of shifting focus and distancing oneself from one's practice, seeks to bring out the various facets of the patient's 'well-being'.

These are the means of practical wisdom. In Ricœur's view, this consists of drawing up a care plan by a movement towards paying careful, plural attention to the patient's narrative and using this to explore how the experiential features of suffering can be linked to the diagnostic and therapeutic data resulting from biomedical analysis. The latter should be structured in a narrative register that enables the patient to restore a forward-looking timeframe, thus allowing them to regain trust in themself, in life and in the care setting that surrounds them.

What matters when dealing with each patient is to go beyond disciplinary expertise, in a deliberative posture that sheds light on the contingent circumstances guiding the choice of the best possible singular action. This choice depends on the health carer's ability to make decisions that guide his actions in the direction of relief of suffering and the patient's recovery.

Like the Aristotelian *phronimos*, the wise, prudent clinician who can find the 'suitable norm' in a singular situation knows, as René Simon says in this connection, that 'the solution is not prescribed, but is risky, uncomfortable, singular, difficult'. According to René Simon, the clinician must:

> in a sense, invent what has to be done, what will be required as a singular duty, the result of a debate that is sometimes difficult and always risky, whose elements do not appear with the clarity that a rational mastery of the situation would ensure; it is in the awkwardness that results from the discrepancy between respect for the norm and respect for people that the decision must be made (Simon 2000, p. 65).[4]

In this context, I believe that the ethical framework for the caregiver relationship, as described by Paul Ricœur, could be improved from a clinical practice perspective by drawing on the thinking of the philosopher Emmanuel Levinas. This suggests a direction for a fundamental shift in focus that seems to me vital to the clinical relationship (Benaroyo 2011).

To what extent could Levinas's thought be relevant in addressing the following questions? How does a healthcare provider create, maintain and ensure the persistence of a trustful climate within the physician-patient relationship? How can they create a common understanding from which an agreement on the benefits being sought through treatment, the means to achieve these benefits, and the roles necessary to realise them, can all emerge? In short, how can he or she create an ethical space from the heart from which one can embark on the restoration of the patient's ability to be?

To me, it seems apparent that the condition *sine qua non* for establishing such a space is to revitalise the original ethical meaning of the clinical relationship, namely to awaken ethical responsibility when faced with the suffering of others. This involves reassessing the ethical status of the original asymmetry which structures the relationship of care and its relationship to the associated aid. The experience of vulnerability related to suffering becomes then a particular source of attention.

Suffering usually drives patients to seek help and to give their trust to a person professionally capable of helping them. Trust initially means placing themselves in a state of dependence and hope, or of extending a state of dependence to a caregiver who promises competent aid and can accomplish it. For the patient, this is an acceptance of vulnerability and an admission that the person to whom their trust is granted may exercise power over them for his or her own sake. It is also believing the caregiver's word and hoping that the powers they have conferred will not be exercised without their knowledge.

From my point of view, the first step towards achieving a space of trust needs to be made by the health carer, in the light of the patient's lived experience of vulnerability. The physician can take the first step towards fulfilling the implied promise of using all possible means to benefit the patient without exerting his power to the patient's detriment by demonstrating to the patient their attendance and their intention to respond to the hope placed in them. Creating a climate of *trust* in medicine is to open a space for dialogue and for commitment to face up responsibly to the patient's suffering.

It is therefore, in my view, relevant at this stage, to refer to the prolific thoughts of Emmanuel Levinas on *otherness*. I believe this provides the essential constituents

[4]The translation of the citation is my own.

for reflection on the theme of trust in the context of the medical care relationship. For Levinas, answering the call of the sufferer,—demonstrating compassion, – and being there for that sufferer are two ethical imperatives at the source of the creation of common humanity resources. This stance opens a space in which common values can be shared and discussed in the context of a clinical care plan.

It is by drawing on the ethical resources of Levinasian concepts of suffering and compassion that it appears henceforth possible to broach, in a fruitful manner, the theme of establishing and maintaining a space of trust. This approach appears to be a necessary prerequisite to the realisation of the medical care objective whilst respecting the patient's values.

3 Emmanuel Levinas' Insight

In one of his first books, *Time and the Other* (Levinas 1985, pp. 55–58), Levinas (1985, pp. 55–56) addresses the theme of suffering:

> The essence of suffering merges with the inability of the sufferer to detach themselves from their suffering. This does not define suffering through suffering but maintains that the implication sui generis constitutes the essence. Suffering imparts no place of sanctuary. It is the fact of being directly exposed to being. It is the fact that there is no possible escape or of going back. All the acuity of suffering is in this impossibility of going back. It is the fact of being driven back both in life and of being. In this sense suffering is the impossibility of nothingness (Levinas 1985, pp. 55–56).

For Levinas, the content of awareness of suffering is 'the unassumable' (Levinas 1998, p. 91). Suffering is in essence 'refractory to the synthesis' (Levinas 1998, p. 91). This lack of conviction, gives suffering a character of 'passivity' (Levinas 1998, p. 91).

However, the character of the 'passivity' of suffering does not mean, according to Levinas, that the latter is accompanied by inertia.: 'The passivity of suffering is made evident by its bewildering and incomprehensible character, in other words, strange in any light, we could say, making it impossible for any assumption of possibility and in which one is entrapped'. (Levinas 1985, p. 58) In this view, 'Suffering defies this intimacy of oneself which throws all our reasoning into disarray' (Levinas 1985, p. 56). According to Levinas (1985, p. 54), it conveys the otherness of our own body, as he highlights:

> Only by arriving at the diminution of their solitude through suffering and their relationship with death, places the sufferer in a situation in which the relationship with otherness becomes possible (…) The relationship with the future is the same as that with otherness (Levinas 1985, p. 54).[5]

The character of the 'passivity' of suffering is thus linked to the notion of sensitivity, specifically, how the subject is affected by otherness and by exposure to

[5] The translation of the citation is my own.

otherness. For Levinas, the sensitivity in suffering explicitly becomes *vulnerability*, i.e. openness to otherness (Levinas 1998, pp. 92–93). 'In suffering, sensitivity is a vulnerability, more passive than receptivity; an encounter more passive than experience. It is precisely an evil' (Levinas 1998, p. 92). Thus for Levinas, the suffering body is the bearer of a fundamental ambiguity: whilst the physical experience of suffering imprisons the subject within themselves, vulnerability confirms that the subject is escaping from themselves, it is in the state of opening up to otherness.

The passivity of suffering can be understood as patience, as an opening to otherness - hence a possible etymological explanation of the word 'patient'. Patience does not convey suffering due to inertia, but rather patience as the possibility of a relationship with otherness, as Levinas points out:

> Is not the evil of suffering—extreme passivity, helplessness, abandonment and solitude—also the unassumable, whence the possibility of a half opening, and, more precisely, the half opening that a moan, a cry, a groan or a sight slips through—the original call for aid, for curative help, help from the other me whose alterity, whose exteriority promises salvation? Original opening toward merciful care, the point at which (...) the anthropological category of the medical, a category that is primordial, irreducible and ethical, imposes itself. For pure suffering, which is intrinsically senseless and condemned to itself with no way out, a beyond appears in the form of the interhuman (Levinas 1998, pp. 92–93).

Levinas tells us that regarding this 'passivity', being face-to-face with the otherness of suffering is the moment of awakening to the responsibility for others. Following Levinas' thought, Catherine Chalier thoroughly describes this call:

> The moment of this upheaval, of this shock, leaving no time for reasoning and with no prospect of escape, confronted with the face of otherness, a face which, through its nakedness and destitution, through its distress which is even greater than mine, and with every chance of mortality, is explicitly calling me, unappealingly tormenting my serenity. It is at this moment that a desire emerges and awakens in me another fear which through its serenity gives me sense of a further anxiety, limiting my horizons. I know I am responsible for the fate of this otherness, responsible for losing, before him, all innocence of my being (Chalier 1987, p. 136).[6]

According to Levinas, this awakening to responsibility, enrobed in the confrontation with the suffering other, is the ethical nucleus of the medical care relationship. Ethical responsibility, in the sense portrayed by Levinas, is thus the moment of hospitality *par excellence.* Compassion-for-others is therefore, in the Levinasian sense, both a questioning prompted by the call of the other combined with a gesture of reception for that sufferer. As Catherine Chalier explains:

> Compassion (...) advocates respect for that which is going on inside the other rather than busily precipitating what is thought to be understood by someone who claims to know. It imposes humility which, when in dialogue with the condemned, "excludes vain consolations and false words of hope". Humility, is difficult for one doing everything possible to alleviate the suffering of another—and diligence here is paramount—but who, at the same time, knows full well that he is not quenching the thirst of the other, the thirst for peace and, above all, the thirst for a sense that will put an end to the nagging question, present in all suffering "Why me?" (...) Hearing these aspirations and being attentive to these expecta-

[6] The translation of the citation is my own.

> tions does not mean that we hold the answers nor does it imply that we possess the comprehension of the meaning of suffering. What it means, is that we must find in ourselves enough willingness to humbly accept the words and silences of another, without immediately suppressing their legitimate volubility and their legitimate concerns. Finally, this necessitates finding in oneself the resources essential to give attention to those who are suffering and that despite the evil or old age which has taken hold of their body, despite the misfortune that mutilates their joy and risks damaging their soul, their singularity, their life, in all its irreplaceable personal uniqueness, is not in vain (Chalier 1987, p. 140).[7]

As we see it, the Levinasian concept of compassion evolves into a dialectic between questioning and availability. With this in mind, we can then see that compassion weaves together the bonds which constitute the ethical core of the physician-patient relationship. Availability appears to me the commanding concept which paves the way for the establishment of common humanity resources within the clinical dialogue: it paves the way for creating a sphere of trust based on the recognition of a common vulnerability, of a "familiar" fragility. By granting ethical status to vulnerability, compassion (in the Levinasian sense), seems to me to constitute the ethical nucleus around which the ethics of responsibility in clinical medicine can be elaborated.

Hence, in my view, a Levinasian reading of the clinical relationship reveals the dimension of the primary moral commitment at the heart of concern for others upon which any clinical relationship must be based (Benaroyo 2022). The latter calls on health carers to perceive ethical responsibility as focusing on the 'visitation' of the patient, rather than on reducing the singular conversation (which leads from diagnosis to treatment and prognosis) to its technical aspects. In other words, this means that health carers' ethical awakening involves accepting their *patients' profound otherness*, which is expressed through their *face* and their associated vulnerability. Here vulnerability is seen not as a loss of autonomy that is to be dealt with as far as possible by scientific medicine following the model of usual medical expertise. Rather, it is seen as an appeal, an ethical injunction to take care of this particular patient by opening up a space in which he or she has a possible future. This conception of responsibility calls for an ability on the part of the health carer to welcome the patient, and to show hospitality. Openness to the patient's *radical otherness* is one of the ways of shifting focus that the caregiver needs to express his vulnerability. It is in this dialogue between two vulnerabilities that trust can start to grow and be deployed in a manner of caring that can respond responsibly to the core of suffering as it is perceived by the patient (Benaroyo 2004).

This approach to clinical care allows room for the *unknowable* at the heart of the patient's experience of existence that it is important to preserve, especially as the medical-technology approach tends to control it. The health carer's attention to the preservation of this unknowable aspect, encouraged by a shift of focus from medical rationality alone, opens up prospects of recognition and trust for the patient, comforting them with the idea that they own their suffering. It is at this point that

---

[7] The translation of the citation is my own.

Levinas' ideas, which are a source of shared humanity, appear relevant to nourish such reflection on the ethical responsibility of healthcare providers.

4 Other Ways of Shifting Focus

Other ways for the 'physician to shift focus from illness to his patient' have been described, such as exposure to works of literature or cinema. Similarly, narrative and hermeneutic approaches have been the subject of publications derived from Paul Ricœur's thinking on narrative identity. I believe that new perspectives on the importance of artistic imagination and experience should be explored to provide various forms of clinical care that can be adapted to the numerous expressions of individual and collective suffering.

5 Perspectives

A conception of medical care based on 'shifts of focus' derived from the thinking of Paul Ricœur and Emmanuel Levinas suggests an archetype whose various facets should be explored in future research. This archetype may take the following ethical steps blazing the trail for practical wisdom (Benaroyo 2022):

- First, an ethical awakening to the vulnerability of the suffering other in his radical otherness—ethics *as* hospitality and love.
- Second, elaborating, based on the first step, a 'pact of care based on trust'—ethics *as* justice and care.
- Third, performing the technical step leading, as far as possible, to healing and curing—ethics *as* social moral norms regulating applied medical sciences.
- Fourth, reaching completion the prudential judgment drawing on practical wisdom that chooses and decides what is the right individualized care in this case—ethics *as* practical wisdom.

This model makes use not only of scientific and technical expertise but also of moral, reflexive and critical skills that lie at the heart of the care relationship. It is in this ethical space of this relationship that the art of interpretation can be embodied to draw up a singular care plan.

I would suggest that these different steps constitute the pillars of a philosophy of medicine grounded in medical practice, which offers the resources needed to construct an ethics of responsible care echoing the ethical requirements unique to clinical medicine. The ethical professional gaze enlightening this philosophy looks at clinical practice and ethics as the two sides of the same coin.

The challenge to medicine today is therefore to tackle head-on both sides inherent in the practice of medicine through an attentive, listening and welcoming ethic that will reveal the various interpretative registers of care. It is by patiently rising to this challenge in the context of a philosophy of care that an ethics of clinical can, in my view, be perfected.

Acknowledgements I would like to express my gratitude to Nicola Banwell, University of Lausanne, for her attention to detail in editing this chapter.

References

Baron, Richard J. 1985. An introduction to medical phenomenology: I can't hear you while I'm listening. *Annals of Internal Medicine* 103: 606–611.

Benaroyo, Lazare. 2004. Soin, confiance et disponibilité: les ressources éthiques de la philosophie d'Emmanuel Levinas. *Ethique et santé* 1: 60–63.

———. 2011. Constructing medical wisdom: the significance of healing narratives for clinical practice. In *Sprache und wissenserwerb. Language and acquisition of knowledge*, ed. Ada Neschke and Hans Rainer Sepp, 233–244. Nordhausen: Traugott Bautz.

———. 2022. The significance of Emmanuel Levinas' ethics of responsibility for medical judgment. *Medicine Health Care and Philosophy* 25: 327–332.

Canguilhem, Georges. 2002 [1968]. Le Statut épistémologique de la médecine. In *Études d'histoire et de philosophie des sciences*. Paris: Vrin.

Chalier, Catherine. 1987. La souffrance d'autrui. In *La persévérance du mal*, 127–143. Paris: Cerf.

Hippocrates. 1999. *L'Art de la médecine. Airs, eaux, lieux*. Trans. J. Jouanna, C. Magdelaine. Paris: Garnier Flammarion.

Jewson, Nicholas D. 1976. The disappearance of the sick-man from medical cosmology, 1170-1870. *Sociology* 10 (2): 225–244.

Lefève, Céline. 2014. De la philosophie de la médecine de Georges Canguilhem à la philosophie du soin médical. *Revue de Métaphysique et de Morale* 82 (2): 197–221.

Levinas, Emmanuel. 1985. *Le temps et l'autre*. Paris: Presses universitaires de France.

———. 1998. Useless Suffering. In *Entre nous. Thinking-of-the-other*, 91–101. Trans. MB Smith and B. Harshav. New York: Columbia University Press.

Merleau-Ponty, Maurice. 1964. *Le Visible et l'invisible*. Paris: Gallimard.

Peabody, Francis W. 1927. The care of the patient. *Journal of the American Medical Association* 88 (12): 877–882.

Reich, Warren T. 1995. *Encyclopedia of bioethics. Revised edition*. Vol. 1. New York. Macmillan.

Ricoeur, Paul. 1992. *Oneself as another*. Trans. K. Blamey. Chicago: Chicago University Press.

———. 2000. Prudential judgment, deontological judgment and reflexive judgment in medical ethics. In *Bioethics and biolaw. Judgment of life*, ed. Peter Kemp, Jacob Rendtorff, and Niels Mattsson Johansen, vol. 1–15, 15–26. Copenhagen: Rhodos International Science and Art Publishers.

Schwartz, Michael A., and Osborne P. Wiggins. 1985. Science, humanism, and the nature of medical practice: A phenomenological view. *Perspectives in Biology and Medicine* 28 (3): 331–361.

———. 1988. Scientific and humanistic medicine: a theory of clinical methods. In *The task of medicine: Dialogue at Wickenburg*, ed. K.L. White, 137–171. The Melon Park: Henry J. Kaiser Foundation.

Simon, René. 2000. La Vertu de prudence ou sagesse pratique. *Ethica* 12: 45–71.

Ten Have, Henk. 1995. The anthropological tradition in the philosophy of medicine. *Theoretical Medicine* 16: 3–14.

Chapter 2
Maternity. The Paradigm of Care in the Doctor–Patient Relationship

Silvia Dadà

Abstract The main purpose of this article is to analyze the category of maternity from a philosophical point of view. This concept has been developed in continental philosophy, especially by Emmanuel Levinas and Adriana Cavarero, who interpret it as an image of ethical subjectivity based on responsibility for the Other. Also, the Ethics of care, developed in the United States since the 1980s, has often used the paradigm of the mother-child relationship to describe the main features of its perspective. Against universal, principled thinking and the impersonal, abstract idea of justice, care thinkers instead value relational, contingent, and particular thinking, which focuses on responsibility towards others, especially the most vulnerable. This paradigm undoubtedly has several critical points, which have been highlighted by care thinkers themselves. These problematic aspects are mainly related to a possible biologistic interpretation that would reduce the woman to a mother and to the only person responsible for care tasks.

However, with these risks in mind, the idea of motherhood can be very useful in counteracting the reduction of inter-human relations to power relations or instrumental/economic relations.

It is particularly in the field of health care that it can be useful in describing the doctor-patient relationship. In this article, I will propose a model of the doctor-patient relationship called "maternalism". This term is rarely used, and when it is evoked, it is often wrongly understood as a form of soft paternalism or, on the contrary, as a total cancellation of the doctor's will. Against these interpretations, I will show the importance of this model, which values neither medical authority alone nor the independence of the patient, but rather their ethical relationship.

The article will be divided into three parts.

1. In the first part, I will try to answer the questions about what maternity as a philosophical category is. I will focus on the main traits of maternity as I interpret them from Emmanuel Levinas and Adriana Cavarero.

S. Dadà (✉)
University of Pisa, Pisa, Italy
e-mail: silvia.dada@unipi.it

J. Braga (ed.), *Ethics of Care*, Advancing Global Bioethics 20,
https://doi.org/10.1007/978-3-031-73608-7_2

2. In the second part I will present the main characteristics of maternal thought in the ethics of care, paying particular attention to the way it distances itself from the contractualist paradigm of homo economicus. The main references will be the thought of Sarah Ruddick, Eva Kittay and Virginia Held.
3. In the third part, I will try to apply this paradigm to the field of medical care, to the relationship between the doctor and the patient. Against the paternalistic relationship, but also beyond the simple protection of autonomy, I will then propose the maternalist model, as an example of care based on the idea of responsibility, trust and empowerment.

Keywords Care · Maternity · Doctor–patient relationship · Levinas · Cavarero · Held · Kittay · Vulnerability

1 Introduction

The concept of "maternity" as a philosophical category has a central role in the philosophy of Emmanuel Levinas, where it is used, especially in his mature works, to describe ethical subjectivity. Although the French philosopher does not expressly speak of "care," he describes the maternal relationship as the ethical focus of the subject, which is exposed to the Other to the point of substitution in responsibility. In the wake of Levinas, but with some differences, Adriana Cavarero also takes up this theme and sees in the maternal inclination the very place to rethink the subject by overcoming the traits of self-referentiality, independence, and domination associated with it, especially in the modern philosophical tradition. These two authors do not simply show the importance of motherhood as if it were one event among others, a possibility that the subject may or may not choose to assume. Rather, they both show that it is precisely through such an idea of maternity that a relational origin of the subject can be thought of, which is ethically structured in its inclination to take care of the Other.

The ethics of care, developed in the United States in the 1980s, has often used the concept of motherhood and the mother–child relationship to describe the main features of its perspective. Like Levinas and Cavarero, they reject the universal, principled concept and the impersonal, abstract idea of justice. Instead, ethics of care thinkers value relational, contingent, and particular thinking, which focuses on responsibility toward others, especially the most vulnerable. Sarah Ruddick in particular speaks of *maternal thinking*, but references to motherhood can also be found in Carol Gilligan, Nel Noddings, Eva Kittay, and Virginia Held.

From these two different traditions, as I will show in this article, it is possible to think of the maternal paradigm as the basis for a philosophy of care that has a solid theoretical framework but also relevant practical consequences and applications. This paradigm has been critiqued on several points that have been highlighted by feminist thinkers and care thinkers. These problematic aspects are mainly related to a possible biologistic interpretation that would reduce the woman to a mother as the only person responsible for care tasks. However, with these risks in mind, the idea of motherhood can be useful in counteracting the reduction of interhuman relations to power relations or instrumental/economic relations.

It is particularly in the field of health care that it can be helpful to describe the doctor–patient relationship. Against the classical paternalistic view, which saw the doctor in a position of absolute dominance over the patient—but also beyond a reduction of the relationship to contractual and formal protection of the patient's autonomy—I will propose a relationship that I will call *maternalism*. This term is rarely used, and when it is evoked, it is often wrongly understood as a form of soft paternalism or, on the contrary, as a total cancellation of the doctor's will. Against these interpretations, I will show the importance of the maternal paradigm, which values neither medical authority alone nor the independence of the patient, but rather their relationship based on care.

This article will be divided into three parts. In the first part, I will try to answer questions about maternity as a philosophical category. I will focus on the main traits of maternity as I interpret them from Emmanuel Levinas and Adriana Cavarero. This first step will allow me to use the concept of maternity in a broader sense than that proposed by the philosophers of the ethics of care, to ground the discourse on a philosophical theoretical framework. I think that the category of care itself is to be regarded as an essential trait of subjectivity, from a perspective such as that proposed by Levinas of "ethics as first philosophy."

In the second part, in proposing the maternal paradigm, I will focus on its main traits starting from the philosophers of the ethics of care. I will pay particular attention to the way it can be distinguished from the contractual paradigm of *Homo economicus*. The main references will be the thought of Sarah Ruddick, Eva Kittay, and Virginia Held.

In the third part, I will try to answer the question of whether the doctor–patient relationship can be interpreted in maternal terms. Starting from the analogy with the maternal paradigm exposed in the first and second parts, I will propose the maternalist model as an example of care based on the ideas of responsibility, trust, and empowerment of the patient.

Before beginning, one more clarification is necessary. I will speak of both "mother" and "mothering person" interchangeably. In both cases, I am not referring to the concrete, biological reality of the phenomenon, but rather to the caring relationship associated with it. Indeed, it is true that motherhood has its foundation on biological grounds, in the natural phenomenon of procreation, birth, and care of the child. Thus, the temptation arises to lead the specific capacity for care back to some "maternal instinct," due to the experiences of pregnancy, childbirth, and breastfeeding (Whitbeck 1984). In its biological sense, however, there is a risk of reducing the woman to her procreative function. As Sarah Ruddick claims:

> Most of us have ambivalent feelings about mothers, including our own; and about being or not-being mothers in a society that either insults maternal work or sentimentalizes it beyond

> recognition while, at the same time, perniciously identifying Womanhood with Motherhood (Ruddick 1983: 4).

There is also the risk of excluding from care men and women who do not have children, whether they cannot or do not want to have them: “If a feminist ethic of care is grounded in these experiences, if the ability fully to care for another is contingent upon the biological experience of mothering, it will necessarily be an exclusive ethic” (Lauritzen 1989: 39). As I am going to show, maternity considered here is a philosophical and ethical category rather than an empirical reality. The maternal relationship in this context is not gender-dependent and therefore I do not interpret it as an exclusively female practice.

2 Levinas and Cavarero: Maternity as a Philosophical and Ethical Category

Although Levinas’s work is extensive and his thought evolved over time (Ferretti 1996), the center of his reflection always remains the ethical relationship with otherness, in the form of responsibility toward the face. The relationship between the Same and the Other is asymmetric and irreducible. The Other invokes the subject and compels him or her to respond, and in the face of this call, this commandment, the Self can only say: “Here I am.” The subject, therefore, cannot escape the relationship with the Other, as it precedes his freedom. The obligation that moves the subject is not dictated by the force or violence of the Other, but by its vulnerability. The Other asks me not to kill him, and in the face of such vulnerability makes it so that I cannot ignore him:

> The face with which the Other turns to me is not reabsorbed in a representation of the face. To hear his destitution which cries out for justice is not to represent an image to oneself, but is to posit oneself as responsible, both as more and as less than the being that presents itself in the face. Less, for the face summons me to my obligations and judges me. The being that presents himself in the face comes from a dimension of height, a dimension of transcendence whereby he can present himself as a stranger without opposing me as obstacle or enemy. More, for my position as *I* consists in being able to respond to this essential destitution of the Other, finding resources for myself. The Other who dominates me in his transcendence is thus the stranger, the widow, and the orphan, to whom I am obligated (Levinas 2007: 215).

The Other cannot be reduced to an object of knowledge. He is encountered as the one who needs my help and at the same time obliges me in the height and majesty of his transcendence. In the relationship, then, the subject loses its ontological and epistemological primacy: the subject, we may say, arises from the call of the Other as one who answers to it. In this responsibility before the face of the Other, then, the subject is no longer for-the-self, but for-the-other: thus, Levinasian subjectivity has the traits of passivity.

There is a shift from the Ego to the Self: Levinas achieves “a radical deconstruction of the subject as master of self and firmly planted in being” (Ciaramelli 1985:

121). A change occurs from a subject-of to a subject-to, and subjectivity becomes *subjection*: "The self is a *sub-jectum*; it is under the weight of the universe, responsible for everything" (Levinas 2011: 116).

The Self as *sub-jectum*, in the accusative form, is vulnerable because of its fragility and the vulnerability of the Other. The subject's responsibility lies precisely in this: he cannot help but respond to the Other's face, in the face of his suffering. Especially in *Otherwise Than Being*, Levinas focuses on the concreteness and corporeality of this passive subject. The vulnerability, the suffering, and the pain become central themes. The relationship between the Self and the Other is not peaceful and abstract, but rather concrete and often painful. In this respect, Gérard Bailhache correctly speaks, for Levinasian thought, of "concrete transcendence" (Bahilache 1994: 246). The Self is exposed to the Other to the point of substitution, so much so that Levinas describes the relationship as "having-the-other-in-one's-skin" (Levinas 2011: 115). The sensitivity and vulnerability of the subject in front of the Other make him restless; he is obsessed with responsibility: "Substitution operates in the entrails of the Self, rending its inwardness, putting its identity out of phase and disrupting its recurrence" (Levinas 2011: 196 n22).

This reference to the entrails returns several times and it has a biblical origin.[1] Levinas states that the expression *rakhem arahemenu*, which means precisely the groaning of the entrails, refers to the sphere of maternity, as *rakhem* means "womb" (Levinas 2006).

It is precisely the image of maternity that is the most successful depiction of ethical subjectivity for Levinas. As Donna Brody rightly points out, "Maternity as subjectivity constitutes Levinas's most sophisticated determination of relation between the Same and the Other—a determination that father–son relation could not alone provide" (Brody 2001: 54). Fatherhood (which Levinas had analyzed in *Totality and Infinity*) is not enough to explain the emergence of responsibility: "It does not do the work of showing how the I [is] 'open' to or always already receptive to the ethical appeal of the face" (Brody 2001: 66). Instead, motherhood succeeds, through bodily sensitivity, in representing the original substitution. As Jennifer Rosato states,

> In Levinas's text, it is generally the physical realities of pregnancy rather than its emotional or psychological effects that are used to represent the discomfort of a self who discovers he is fundamentally constituted as a "being-for-the-other"; this helps to emphasize the fact that the encounter with an Other is meant to be a sort of physical sensibility (Rosato 2012: 352).

This notion makes it possible to describe the subject as involved in the pre-original relationship with the Other, as responsible to the point of substitution, and as materially and concretely involved in the relationship: "The subjectivity of sensibility, taken as incarnation, is an abandon without return, maternity, a body suffering for another, the body as passivity and renouncement, a pure undergoing" (Levinas 2011: 79). So, the subject, like a mother, carries the Other within himself.

---

[1] *Jeremiah* 31:20: "Is Ephraim a dear son unto me? is he a child of delights? For whilst I have been speaking against him, I do constantly remember him still. Therefore my *bowels* are troubled for him: I will certainly have mercy upon him, saith Jehovah" (Darby translation, my emphasis).

> In maternity what signifies is a responsibility for others, to the point of substitution for others and suffering both from the effect of persecution and from the persecuting itself in which the persecutor sinks. Maternity, which is bearing par excellence, bears even responsibility for the persecuting by the persecutor. Rather than a nature, earlier than nature, immediacy is this vulnerability, this maternity, this pre-birth or pre-nature in which the sensibility belongs (Levinas 2011: 75).

Maternity is a form of patient waiting: the subject has the Other in him, carries him, and supports him. It is also total participation in the Other's pain: I feel with the Other, and I suffer for him. For Levinas this is not a gender difference nor a sexual difference (Dadà 2021): it is the subject as a human that is maternal: "Motherhood means responsibility for others" (Levinas 2011: 94).

This idea of maternity is understood here as a phenomenon before birth, involving the subject in an original relationality. This is different from, for example, Adriana Cavarero's perspective, which resorts to motherhood as an image of the post-birth care relationship. Echoing Arendt's view, Cavarero sees the human condition in terms of natality, which is the concrete beginning of existence, the carnal openness to performative action. In birth, therefore, the relationship with the world and with others is realized in the form of the mother–child relationship.

Also in Cavarero's case, as with Levinas, the subject is revisited in a new ethical and ontological dimension. Western philosophy has always described the subject as a vertical (and masculine) entity. Self-centered, and epistemologically privileged, the subject finds autonomy and independence as its most proper characterization. Even from an ethical point of view, its virtuous character is associated with righteousness, against all forms of "perversion." In contrast, the idea of inclination has been interpreted negatively:

> The aforementioned "upright man" referred to in the tradition, rather than an abused metaphor is literally a subject who adheres to the uprightness of the straight axis that serves as the principle and norm of his ethical posture (Cavarero 2013: 14).

Inclinations, passions, and feelings that move the soul were the enemy of Enlightenment-era rationalism and are still regarded controversially and ambiguously today. Cavarero, however, takes up the idea of the inclination to rehabilitate the more passionate and relational dimension of subjectivity. Through the analysis of Leonardo da Vinci's painting the *Virgin and Child with Saint Anne*,[2] Cavarero attributes this ontological and ethical posture to the woman and in particular to the mother. In the painting we can see the Virgin leaning toward Jesus: this gesture represents the care for the child, which is the symbol of vulnerability. Another mother is present, Saint Anne, who also turned toward Mary. This triangulation of the characters perfectly describes the relational trait of motherhood and the bond that joins the subjects in care and responsibility toward the vulnerability of the weakest:

---

[2] For the painting see: http://en.wikipedia.org/wiki/The_Virgin_and_Child_with_St._Anne_(Leonardo_da_Vinci)

> Generated by an endless chain of other mothers lost in the night of humanity, she gives birth to vulnerable beings, and, above all, she has the not at all symbolic role of the one who is responsible *for* them, beyond the act of procreation (Cavarero 2013: 144).

So, while the phenomenon of motherhood is approached by the two authors in different ways[3] and at different times (Levinas as gestation, Cavarero as a post-birth relationship), it is worth holding these two perspectives together to think of a philosophical paradigm of maternity. A joint reading allows us to interpret subjectivity in relational, concrete terms, valuing care and responsibility for the Other as essential and original aspects of human existence.

With these premises and philosophical foundations in mind, I will now consider the ethical perspective of maternal care.

3 Maternal Thinking in the Ethics of Care

From the perspective of the ethics of care, since the 1980s, mothering was an example of a caring relationship par excellence. For example, Carol Gilligan in her book *In a Different Voice* defines care as "maternal" on several occasions.[4] To introduce the theme of care, Gilligan uses the myth of Persephone, who talks about the mother–daughter relationship and the importance of this relationship for the maintenance of the balance of life:

> The myth of Persephone speaks directly to the distortion in this view by reminding us that narcissism leads to death, that the fertility of the earth is in some mysterious way tied to the continuation of the mother–daughter relationship, and that the life cycle itself arises from an alternation between the world of women and that of men (Gilligan 2003: 23).

The way in which mothers act and think questions the principles on which morality has always been based—namely autonomy, independence, and the impersonal universality of the law. It is rather a vulnerability and dependence that become central and activate a network of relationships that make up our society:

> Although Freud, claiming that "we are never so defenseless against suffering as when we love…" pursues the line of defense as it leads through anger and conscience to civilization and guilt, the more interesting question would seem to be why the mother is willing to take the risk. Since for her love also creates the possibility of disappointment and loss, the answer would seem to lie in a different experience of connection and a different mode of response (Gilligan 2003: 28).

Also, Nel Noddings defines caring as the "language of the mother" as opposed to the "language of the father," namely justice:

---

[3] Cavarero criticizes Levinas' perspective on the topic of motherhood, saying that "it is disappointing and discouraging" (Cavarero 2013: 140).

[4] For exemple, see Gilligan (2003: 74).

> The traditional approach, that of the father, is to ask under what principle the case falls. But the mother may wish to ask more about the culprit and his victims. She may begin by thinking, "What if this were my child?" (Noddings 2013: 36).

Sarah Ruddick describes motherhood as "motherwork," entailing a union of theory and practice. She interprets the maternal relationship as the epistemological connection between practice and thought, action and knowledge, where thoughts are determined by actions and vice versa. Indeed, motherwork is a relationship in which the mother responds to her child's needs through works of what she calls "preservative love, nurturance, and training" (Ruddick 1990: 17). This engagement, in turn, gives rise to maternal thinking, which is a cluster of metaphysical attitudes, cognitive capacities, and values. Hence, maternal thought is in constant negotiation between the abstract and the contingent, the particular and the universal. Ruddick says it requires a continuous search for balance in the double dimension of the large and the small, the immediate and the eternal.

This type of judgment is not only the result of the rational faculty but also of the affective and emotional one. Without falling into sentimentalism, or irrationalism, Ruddick argues that mothering requires the "intelligence of emotions," as Martha Nussbaum (2022) calls it. For as much as motherwork and maternal thinking seem to be completely guided by empathic feelings and emotions, they are also the result of careful thought, judgment, and evaluation. Moreover, the emotions of mothering are not only positive: the mother, in caring, also feels fear, anger, and sometimes even hatred. This ambivalence is a hallmark of the mothering practice. Although mothering is not considered to be intrinsically peaceful, it is, in Ruddick's opinion, "a 'natural resource' for peace politics," because the mother "maintains conditions of peacefulness so that her children can grow in safety" (Ruddick 1990: 160). Indeed, mothers struggle daily with finding and using nonviolent strategies, reconciling relationships, and keeping the peace.

Eva Kittay, in her book *Love's Labor*, analyses Ruddick's thinking and enriches some concepts from the experience of being a mother of a disabled child. In ordinary cases, the child is a vulnerable and dependent being who achieves his or her independence by growing up with the help of the mother. By contrast, caring for a disabled child shows us that this autonomy often cannot be achieved:

> So when we think of mothering of a disabled child as enabling and fostering development, we must also reconceive development, not only toward independence, but toward whatever capacities are there to be developed. Development for Sesha means enhancement of her capacity to experience joy (Kittay 1999: 173).

Kittay argues that autonomy is an ideal that is often unattainable and that equality can rather be thought of as starting with dependence. This is the meaning of the expression "We are all some mother's child" (Kittay 1999: 115), which means that we are all the same because we are all dependent on someone else's care.

This dependence is related to trust, another central theme of the ethics of care. Annette Baier addresses this issue, making a careful examination of the various forms of trust and the risks it entails. Trust, according to Baier, is about accepting vulnerability, for when we trust someone or entrust something to someone, we open up the possibility of losing something important and being betrayed:

> Trust is an alternative to vigilance and reliance on the threat of sanctions; trustworthiness is an alternative to constant watching to see what one can and cannot get away with, to recurrent recalculation of costs and benefits. Trust is accepted vulnerability to another's power to harm one, a power inseparable from the power to look after some aspect of one's good (Baier 1994: 133).

In the wake of Hume's thought, Baier considers trust as based on original sympathy, or empathy, which allows us to feel the emotions of others, especially those closest to us. The affective dimension is complemented by a cognitive dimension, which makes trust a guide to moral action (Baier 1987). The mother-and-child relationship best represents the deepest level of trust: the child relies completely on the mother without the need for rational justification. As the child grows, the dynamics of the relationship and the mode of trust also change. Thus, trust moves from being absolute reliance to thoughtful trust. Although blind trust has, for Baier, no moral value, nevertheless it is the condition of possibility without which other forms of trust would not be possible.

Mothering, according to these thinkers, represents the caring relationship par excellence. Virginia Held, in my opinion, sums up the main features of the mothering model most clearly. In her article entitled "Non-Contractual Society: A Feminist View" (1987), she argues that mothering offers us a different model from contractual theories. Held prefers to speak of "mothering person" rather than "mother," to emphasize the fact that the mothering subject can be either a man or a woman. Even in the contractual model, we speak of "rational contractors" in the same neutral sense.

The notion of a maternal perspective does not mean the return to a pre-contractual society, such as the feudal model based on family and legacy; rather it takes a post-patriarchal perspective of motherhood that can serve as a model for other social relations. It is not a matter of eliminating or completely replacing the contractual model: for Held, it is necessary to propose an alternative to it, to deal with the variety of relationships in different ways. In fact, we are used to interpreting all relationships and subjects in these economic terms, but this is not the only possibility: "Of course, we can theoretically treat them *as* contractual, but *should* we?" (Held 1987: 115).

Not all relationships are reducible to a contract: that between mother and child, for example, is profoundly different. First, in the case of contractualism, society is thought of as a collection of individuals with self-interests: in this perspective, there is no room for cooperation and trust. In the case of the maternal model, on the other hand, society is a network that rests on precisely these aspects and the interdependence of people.[5]

---

[5] As Annette Baier states, we are "second persons," because "in learning from other persons, we acquire a sense of our place in a series of persons, to some of whom we have special responsibilities. We acquire a sense of ourselves as occupying a place in an historical and social order of per-

Furthermore, the two models describe the subject very differently: the notion of *Homo economicus* is in fact solipsistic and driven only by the narcissistic desire for self-assertion and competition, while the mothering person focuses on care and responsibility for others.

The author then analyzes the main features of the mothering person—child relationship. The first aspect that is emphasized is the *involuntariness* of the maternal relationship. Indeed, motherhood is not a contract, and the mothering person cannot choose who will be her or his child; and conversely, the child cannot choose his/her mother. Maternal care is not really a choice: for we cannot choose to be or not be interdependent, and once the vulnerable Other (the child) is in the world, there is no getting out of the relationship.

The second important point is the asymmetry and non-reciprocity of the relationship. This does not mean that there is absolutely no reciprocity of care between mother and child, but rather that this reciprocity is not governed by forced obligations. Maternal care, in fact, escapes the circular cost–benefit logic. Rather than the economic model of exchange and reward, care is better suited to that of gifting.[6] Care is characterized by a "logic of over-abundance, *surplus* and *excess*," which is the logic of love (Ricoeur 1991), in which we are constitutively, ontologically *indebted* to each other.

The third aspect that Held emphasizes is the uniqueness and permanence of the relationship. For *Homo economicus*, everything has a price and can be traded, while every mothering person is unrepeatable, as is every child. This idea of irreplaceability allows for a distinction between the individualism of interests and the uniqueness of each person inscribed in society.

Furthermore, unlike the contractual model, equality is not based on rights, but on common dependency. In the contractual model, equality is usually associated with the realm of rights. Of course, this dimension of equality is also important, but according to Held it is not the only one:

> Parents and children should not have equal rights in the sense that what they are entitled to decide or to do or to have should be the same. A family of several small children, an adult or two, and an aged parent should not, for instance, make its decisions by majority vote in most cases. But every member of a family is worthy of equal respect and consideration. Each person in a family is as important as a person as every other (Held 1987: 128).

The idea of freedom also changes in the mothering relationship and no longer indicates non-interference. For if the mother respected the freedom of others simply through non-interference, this would lead to the abandonment of the vulnerable child. In the maternal relationship, as well as in many other contexts, the idea of a "Robinson Crusoe" subject is inadequate:

> This Robinson Crusoe image of "economic man" is false for almost everyone, but it is totally and obviously false in the case of infants and children, and recognizing this can be

---

sons, each of whom has a personal history interwoven with the history of a community" (Baier 1994: 87).

[6] The relationship between care and gift was analyzed by Italian philosopher Elena Pulcini (2009).

> salutary. It can lead us to see very vividly how unsatisfactory are those prevalent political views according to which we fulfill our obligations merely by refraining from interference (Held 1987: 129).

The relationship between individuality and sociality is thus opposite. Unlike the social contract, in which there is the creation of a community from autonomous and separate individuals, maternal care takes place as a progressive loosening of the bond with a view to the creation of a personality, that of the child, unique and irreplaceable. In turn, this subjectivity will open, over time, to other relationships of care and responsibility.

Another aspect in which the maternal model differs from the contractual model is its conception of power. On the one hand, power is seen as a means of prevailing and subjugating others to one's own power, whereas in the case of the mother, power is used for *empowering* the child. The mothering person is extremely powerful in front of her/his child, but she is also maximally exposed to her child's demands. The child's power, therefore, consists precisely of his vulnerability—as in the case of Levinasian maternity, in which the mother is vulnerable *for* the vulnerability of the other. So, power is understood in a non-conflictual sense; that is, as the development of possibilities for the child. Although there are cases in which the maternal relationship becomes one of domination, the exercise of force, and even violence, this does not refute Held's argument. This possibility simply reveals the failure of the relationship and the fact that it is not a true maternal caring relationship: "Of course the relation may be degenerate from being one of domination and submission, but this only indicates that the relationship is not what it should be" (Held 1987: 131).

So, these are the main aspects of the maternal paradigm. From a theoretical point of view, it is based on an idea of the relational subject, which has the inclination toward the vulnerable Other as its essential trait. Starting from this philosophical approach, as just shown, the idea of mothering can be very useful in counteracting the reduction of inter-human relations to power relations or instrumental/economic relations.

4 Maternalism in the Doctor–Patient Relationship

The caring relationship, exemplarily represented by the maternal paradigm, has been considered in the previous paragraphs in its philosophical and ethical aspects. I will now show how this maternal perspective can also be useful in thinking differently about the doctor–patient relationship.[7] I will then start with an examination of

[7] Although other relationships are also very important (e.g., between patient and nurse, between patient and caregiver, and between patient and family members), I decided to consider primarily the one with the physician for two reasons. First, because the critical debate on the topic is very rich. Second, that relationship is the one that has run the greatest risk of a reductionist interpretation, undermining the true nature of care. The analyses in this section can also be applied, with some specific variations, to other types of relationships in the field of health care.

the main ways in which the therapeutic relationship has changed throughout the history of medicine. In particular, I will focus on paternalism and the model based on the principle of autonomy. I will try to highlight their weaknesses and strengths, and then I will propose an alternative model called *maternalism*. This model, based on the maternal paradigm of care, allows for avoiding both the risk of total subordination of the patient to the doctor, and the danger of abandonment in the name of defending self-determination. It especially allows the clinical relationship to be interpreted as a *caring* relationship, and not as a power relationship or contract.

Many scholars have studied the clinical relationship and have identified various models to describe it.[8] Despite different names being used, we find references to paternalism in almost all proposals. From an etymological point of view, the term comes from the Latin *pater,* "father." Such reference to the paternal role indicates not only the affective dimension in the family bond but also the character of authority and the exercise of power. The *pater familias* in Roman law was precisely the one who exercised *patria potestas* over material goods, slaves, his wife, and his children: he could decide for them, even depriving them of their status as persons.[9] The term "paternalism" first appeared in the nineteenth century, and denotes, in a negative sense, those policies in which the authority of the rulers, while acting on behalf of the ruled, do not consider them capable of participating in the decisions. The distinctive feature is precisely this coexistence of the authority over others and its benevolent, or presumed benevolent, purpose.[10] Even in the medical context, we speak of "paternalism" in a similar sense. As is well known, for a long time in the history of medicine, the paternalistic model prevailed. In this model, the doctor assumes that he knows what is best for the patient, better than the patient himself, based on his medical knowledge. He then partially provides information to obtain consent to what he thinks is best. This model is strongly asymmetrical and sees in the doctor an almost unquestionable authority that the patient must simply rely on. There are some cases where this model is necessary, such as in emergency cases of patients with acute trauma, coma, or delirium, where consent cannot be obtained without endangering the patient's life. In other cases, where the patient is conscious and capable, the paternalistic perspective becomes more problematic. I will focus only on this second sense, called *strong paternalism*. As Beauchamp states:

---

[8] For example, Mark Siegler (1984) proposes a chronological conceptualization of the different ages of medicine and the patterns of relationships prevailing in it. He calls them "the Age of Paternalism," "the Age of Autonomy," "the Age of Beaurocratic Parsimony" and finally "the Age of Shared Decision-Making." To take another example, Ezekiel Emanuel and Linda Emanuel (1992) talk about four models in a non-chronological sense: the paternalistic model, the informative, the interpretive, and the deliberative. These two perspectives are both descriptive and prescriptive. In fact, they show what the relationship is like but also express an opinion on how it should be. In the case of Emmanuel and Emmanuel, for example, their favorite model is the deliberative one. While both are valid perspectives, I prefer to follow Emmanuel and Emmanuel's because they propose ideal models, and thus they transcend the historical dimension.

[9] For a philosophical–historical analysis of the idea of *person* see Esposito (2015).

[10] On this concept, see the entry "Paternalism" in the Stanford Encyclopedia of Philosophy (https://plato.stanford.edu/entries/paternalism/#WeakVsStroPate)

> Paternalism is the intentional overriding of one person's autonomous choices or actions by another person, where the person who overrides justifies the action by appeal to the goal of benefiting or of preventing or mitigating harm to the person whose choices or actions are overridden. A person's choices or actions must be substantially autonomous to qualify as paternalistic under this definition (Beauchamp 2009: 81).

In the paternalistic model, the physician provides information in a biased or tendentious manner so as to induce the patient to accept his treatment proposals. In some cases, he even proceeds without obtaining the patient's explicit consent. So, in this model, the principle of benevolence prevails over that of autonomy, while the patient's needs take precedence over his rights.

The good of the patient is thought of as an objective and neutral value, but in a pluralistic society like ours it could happen that the conceptions of the good held by a doctor and by a patient are not the same. In fact, they may diverge both regarding the *meaning* of health and disease and their *value.* This perspective, therefore, justifies the prevalence of the physician over the patient because of the objectivity of values and the epistemological asymmetry of the relationship.

In an attempt to overcome the paternalistic subjugation of the patient to the will of the doctor, especially since the 1970s, particular attention has been paid to the principle of autonomy, which has become the main focus of bioethics (Wolpe 1998).

The defense of the patient's decision-making autonomy comes through the reduction of the doctor's role. In this model, called also the *informative model* (Emmanuel and Emmanuel 1992) the physician is seen as a technician who provides all the scientific information to the patient without intervening in the decision. There is a clear distinction between facts and values: the physician provides information about facts, but does not give opinions about values.

Informed consent is the tool by which this model is realized (Faden and Beauchamp 1986). The patient signs a document declaring that he has received the information and takes responsibility for his choices. Informed consent is thus a means of defending the patient's rights and protecting physicians from possible legal action taken against them. The active involvement of the patient through informed consent is a step forward from the paternalistic approach, in which the physician was free to choose how, if, and when to make the patient participate in the diagnosis and prognosis. The information provided by the physician must in fact be clear and truthful, and expressed in a way that is personalized to the patient's capabilities. Consent, then, must be expressed explicitly, without being delegated to a third party (except in special cases). In this model, the physician and the patient are regarded as *rational contractors.* Robert Veatch (1972) is one of the strongest proponents of this model, as he argues that the contract is the only way to ensure respect for the autonomy and the shared responsibility of each member of the agreement.

This model presents some critical points. In particular, autonomy can be interpreted in terms of non-interference by the patient and independence from the doctor (Manson and O'Neill 2007). The clinical relationship, therefore, runs the risk of no longer being a therapeutic alliance based on trust. This model does not require the physician to take a caring approach, but rather to competently perform the interventions chosen by the patient: for this reason, is a model that promotes the depersonalization of medicine. Then the informed consent risks being reduced to a mere formal contract. This perspective also has a static idea of values, and does not admit that some values may arise from confrontation and process of shared decision-making: "The informative model's conception of autonomy is incompatible with a vision of autonomy that incorporates second-order desires" (Emanuel and Emanuel 1992: 2224).

Both the paternalistic and autonomy-based models undermine the caring relationship. In different ways, in fact, the patient is alone—in the first case because he is subjected to the authority of the doctor who takes for granted what his wishes, aspirations, and desires are. In the case of autonomy, on the other hand, the relationship is often compromised by the excessive formality of communication and the lack of real dialogue around the perspectives of the sick person. The patient is in fact alone in front of a reality that, as much as he understands, he may not be able to cope with.

The covenant of trust between doctor and patient is broken in both cases: on the one hand, in the paternalistic model the doctor replaces the patient, and thus excludes him in decisions; on the other hand, in the autonomy model, the patient becomes defensive and sees the doctor's opinion as interference. The signing of a contract, as in the case of informed consent, establishes everyone's responsibilities and ensures the fairness of the procedure. However, although it may seem like a confidence-saving solution, it alone is not enough. On the contrary, it can undermine what it seems to protect. As Alfred Tauber rightly says, "Law appears when trust fails" (Tauber 2005: 185).

So, against the classical paternalistic view, which saw the doctor in a position of absolute dominance over the patient, but also *beyond* a reduction of the relationship to contractual and formal protection of the patient's autonomy, I want now to analyze the model that I call *maternalism*, which is similar to Emanuel and Emmanuel's deliberative model but differs in some aspects.

The term "maternalism" is rarely used, and when it is evoked, it is often wrongly understood as a form of soft paternalism or, on the contrary, as a total cancellation of the doctor's will. In particular, Laura Specker Sullivan (2016) speaks of "medical maternalism" concerning communications of cancer diagnosis in Japan. In this case, however, the model maintains a strong emphasis on the physician's freedom of decision-making over the patient's wishes: "I define maternalism as deciding to act in an autonomous person's best interests and likely in line with that person's will, but in the absence of the affected individual's expression of consent or assent" (Sullivan 2016: 442). Of course, deciding for someone does not necessarily mean overriding his or her wishes: rather, it may mean interpreting and enacting them. For Sullivan, this work of interpretation is permitted only within a long-term relationship in which the bond is built on the specificity and uniqueness of the patient:

> Paternalism, based on one individual's disregard for another's will, focuses conceptual attention on a general form of relationship that is at once relationship of anyone and no one. Maternalism, by contrast, is predicated on the exercise of relationships in which one party can discern the will of another without explicit communication (Sullivan 2016: 439).

Sullivan's perspective has the merit of highlighting the difference between choosing something *for* someone and choosing something *instead of* someone. However, this model seems to run the risk of abuse by the physician in a paternalistic way.

Mary Mahowald (1987), describes maternalism as a model of a doctor–patient relationship very close to what I have described in terms of autonomy. She interprets this relationship as the total withdrawal of the mother/physician in favor of the patient's independence.[11]

> If we use the term [maternalism] in accordance with its etymological meaning (from the Latin *mater* or mother), it means the opposite of paternalism. To mother is to give birth, and/or to nurture another individual. Giving birth essentially entails letting go of what has been a part of oneself. Nurturance means helping another to be whoever he or she is, independently of oneself. To be maternalistic, therefore, implies respect for the other's (e.g., child's or patient's) autonomy, even at the expense of the nurturer's autonomy, and even at some risk to the other (cf. the newborn who must immediately breathe on her own). The very term "patient" seems inappropriate for this context: "client" may be better, or possibly "consumer" (of health services) (Mahowald 1987: 24).

Mahowald's view reveals some critical points. First, being a mother and being a father are not essentially opposites. Mahowald presupposes certain stereotypes that she claims to criticize. Another of these is the total annihilation of the mother's wishes and desires in favor of the child and thus, in analogy, the reduction of the role of the physician in favor of the role of the patient. The idea of maternalism that I propose, on the other hand, is different from both of these approaches, for is based on those aspects that I have shown in the previous paragraphs.

First of all, it must be said that the maternalistic model is a relational model, whose objective is not to achieve the patient's independence from the doctor but to create an effective caring relationship aimed at his or her health. As Levinas, Cavarero, and writers on the ethics of care have shown, the development of the individual often does not coincide with autonomy, and we can often say that in the medical field, health does not coincide with individual autonomy or the absence of vulnerability of the patient. Illness is in fact one of the experiences that most reveal to us the illusory nature of individual independence from others and obliges us to mutually recognize our vulnerabilities.

The patient is seen as the one who asks the physician for help in confronting the disease. The physician, in turn, responds to the patient on behalf of his own inclination. The individuals involved in the therapeutic relationship are both vulnerable and dependent and are united by an ethical bond that cannot be ignored. Like a mother, the physician takes responsibility for the patient himself. This does not mean that the patient should remain completely dependent. Rather, it means that it is the doctor's ethical duty to take care of him and not abandon him and respect his will and freedom.

Secondly, the clinical relationship, like that of mothering, is constituted by the union of theory and practice, rationality and affectivity, and attention to universal knowledge and the particular situation. A good mother takes care of her children's

[11] What I am going to illustrate is instead closer to what the author calls "parentalism."

needs, both material and emotional. In the same way, a doctor, to be a good doctor, must combine *care* and *cure*, emotional capacity, and scientific expertise.

In fact, the physician's task is not only related to his scientific knowledge, so providing his techniques and information is not enough. Lazare Benaroyo argues that the thought of Levinas and Ricoeur can be useful in this regard for the clinical relationship.:

> Hence, in the realm of clinical medicine we are now facing the *paradox* that lies at the root of the unfolding of ethical responsibility in the sense that Levinas gives to this word. The responsible healer has a double face of Janus: on the one side, the naturalistic rational attitude, to recall Husserl's terms, is necessary to observe the patient with objectivity in order to master his disease processes, and on the other side, the personal pre-rational attitude entails hospitality to welcome the other's vulnerability as an ethical calling. It is out of a synthesis of both opposite attitudes—which respectively call on radically different goals—that responsibility-for-the-other, guided by practical wisdom in the vein of Ricoeur's thought, may be effected in clinical care (Benaroyo 2022: 331).

Thus, ethical responsibility and care are the basis of trust in the clinical relationship. The mother does not claim to know her child's good independently of his will. She helps him discover himself and his needs; she exercises her responsibility by interpreting her child's requests and actively promoting them. Trust between doctor and patient, in a similar way, is made possible by a relationship of transparency. Trust changes as the relationship changes: initially, it is a blind reliance on medical care, then, by increasing awareness and trust, becomes awareness, on the bases of confrontation and sincere dialogue.

With the above considerations in mind, according to Held's perspective, we can also see that although the clinical relationship has sometimes been interpreted in contractual terms, in my opinion it cannot be reduced to this model, for many reasons.

First, the relationship between doctor and patient is not voluntary. Of course, becoming a doctor is a free choice: but once this choice is made, the physician is ethically obliged to render aid and care to the sick person, no matter who the patient is. Care prescinds from the qualities of the individual: as for Levinas, in the clinical relationship, the Other is Other in an absolute sense.

In addition, according to the maternalistic perspective, the relationship is not based on reciprocity. Although the doctor does not give a free service, the purpose of their relationship is not the equivalence of economic transition. The outcome of the work of care, for the physician, is not merely the salary, but the health of the patient. This health is not measurable in simple quantitative terms or as the elimination of the disease. In essence it is a gift, which is achieved thanks to an ethical relationship of commitment and mutual respect.

Unlike a contractual model, the clinical relationship cannot be formalized: it is always unique. Each relationship is different from the others, even if the disease is the same. As in the case of mothering, each child is different and the mother takes original care of each one.

Some scholars think that the best metaphor to describe the doctor–patient relationship is friendship, or "medical philia" (Entralgo 1969; Fried 1974). This perspective allows aspects of paternalism, such as care and love, to be held together with aspects of the autonomy model, such as respect and equality. Friendship, as well as mothering, in fact, is also a unique caring relationship that escapes economic logic and the contractual relationship. The metaphor of friendship is problematic in some respects, in that clinical relationship is not symmetrical as it should be between friends. The doctor–patient relationship is based on three levels of asymmetry: knowledge, responsibility, and emotional involvement. Indeed, the physician has more technical and scientific knowledge than the patient, which makes him as responsible for the patient's care as the patient himself. We might even say that is *more* responsible than the patient: in fact, he has both the responsibility to cure the patient and to inform him sufficiently about his desease condition. The patient, on the other hand, is emotionally involved to a greater extent than the physician and is also the only one physically affected by the disease.

However, these forms of asymmetry do not imply a lack of equality. In fact, equality becomes a relational concept in that everyone is dependent on each other in the relationship, whether it is the mother with the child or the doctor with the patient. As Virginia Held states for mothers and children, in the same way, physicians and patients should not have equal rights in the sense that what they are entitled to decide should be the same. But every member of the clinical relationship, as in the family, is worthy of equal respect and consideration.

Autonomy retains an important role in this model but is understood differently. In fact, being autonomous does not mean, as in the contractual model, being independent. Rather, being autonomous means making decisions consciously within a network of relationships. In this regard, it is correct to rethink autonomy in relational terms.

The concept of relational autonomy has been explored in recent years on philosophical and political (Mackenzie and Stoljar 2000) but especially bioethical grounds (Sherwin 2008; Verkek 2001). In fact, some scholars consider the last 20 years as the "relational turn of bioethics" (Jennings 2016). The patient is not conceivable outside of his ties, and the exercise of his decision-making capacity depends on them. It is impossible to disregard this network in reconstructing the individual's decision-making process:

> Because most of us simply do not envisage ourselves as independent decision- making machines with inviolable rights to access or control "our" data, tissue or bodies alone, especially when that data, tissue, and even our body has real and symbolic meaning for others, autonomy is most useful as an ethical norm when we recognize that it does not mean simply being left alone to decide. For it is often in these messy grey zones of in-betweenness and hybridity where the impact of our decisions on others is not only contemplated but valued, that the truest expression of self-rule is manifest (Dove et al. 2017: 162).

Besides being unrealistic, exclusive autonomy is also undesirable, as it would hinder the patient from achieving self-determination. Especially when it comes to the affective or medical context, the opinion and support of others become a key factor in the patient's autonomy. As in the case of the mother, the patient should not be left alone in the clinical context. As John Hardwing states: "I am a husband, a father, and still a son, and no one would argue that I should or even responsibly could take a sabbatical, another job or even a weekend trip *solely* on the basis of this is what I want for myself" (Hardwing 1990: 6).

Thus trust, care, asymmetry, mutual dependence, and autonomy in the relational sense characterize the maternalistic model of the doctor–patient relationship.

A final fundamental aspect concerns the notion of power. In a caring relationship, asymmetry and interdependence do not mean submission, because the relationship is not based on domination and exercise of power.

As in the case of mothering, also in the doctor–patient relationship power is understood differently than in the contractual model: not as dominance, but rather as the ability to *empower* the patient. Here power is seen not so much with its connection to the force, but rather as the Aristotelian *potentiality*, in its connection to the realm of possibility.

Both in the case of paternalism and in the autonomy model, the relationship between doctor and patient is conflictual, since in paternalism the doctor's will prevails, and in the autonomy model only the will of the patient is considered. In maternalism, on the other hand, as in mothering, the physician helps the patient to decipher the meaning of his or her needs. Patient empowerment consists precisely of the relational ability to create shared values (Tang et al. 2010). Indeed, the patient's needs must be considered together with the doctor's opinions and values, to help the patient see his experience from another perspective. The physician, in this way, provides the patient with the tools to learn about new possibilities for the activation of the patient's latent resources (Palumbo 2017).

Mother and child are not adversaries or enemies, but allies, as are the doctor and the patient. The difference in roles is necessary and does not pose a danger to their relationship. This difference serves rather to enrich each other and create shared values.

5 Conclusions

Maternity is a metaphor that succeeds in fruitfully describing human relationships based on care.

In our age, there is an increasing tendency to regard others as strangers because of moral pluralism (Siegler and Childress 1984), but this attitude destroys the relationship of trust. The contract between autonomous and rational individuals may seem a viable alternative to trusting others, but actually it is not. It is important to change perspective by remembering those aspects that make us all equal: our vulnerability, and the fact that we all need the care of someone else. As Eva Kittay states, "We are all some other's child."

As I've shown in this article, it is precisely the maternal paradigm that encourages this change of perspective. It is particularly useful in the medical field to describe the doctor–patient relationship.

In fact, the maternalistic model makes it possible to overcome paternalistic submission to the doctor without renouncing the relationship of trust. At the same time, it makes it possible to protect autonomy, but in a relational, not individualistic sense.

The doctor is not a guardian, as in the case of paternalism; and he is not an informer, as in the autonomy model. Nor is he a friend, as some scholars argue. The physician is rather a mother, who builds a relationship with the patient based on trust, responsibility, and care.

References

Bahilanche, Gérard. 1994. *Le sujet chez Emmanuel Levinas*. Paris: Puf.

Baier, Annette. 1987. *Postures of the mind: Essays on mind and morals*. Minneapolis: University of Minnesota Press.

———. 1994. *Moral prejudices. Essays on ethics*. Cambridge: Harvard University Press.

Beauchamp, Tom L. 2009. The concept of paternalism in biomedical ethics. *Jahrbuch für Wissenschaft und Ethik* 4: 77–92. https://doi.org/10.1515/9783110208856.77.

Benaroyo, Lazare. 2022. The significance of Emmanuel Levinas' ethics of responsibility for medical judgment. *Medicine, Health Care and Philosophy* 25: 327–332. https://doi.org/10.1007/s11019-022-10077-0.

Brody, D. 2001. Levinas's maternal method from "time and the other" and the otherwise than being. No woman's land? In *Feminist interpretation of Emmanuel Levinas*, ed. Tina Chanter. University Park: Pennsylvania University Press.

Cavarero, Adriana. 2013. *Inclinazioni: Critica della rettitudine*. Milano: Cortina.

Ciaramelli, Fabio. 1985. Dal soggetto alla soggezione. *Aut Aut Firenze* 209–210: 117–128.

Dadà, Silvia. 2021. Levinas e il femminile. Tra stereotipo e d etica. *Etica e Political (Ethics and Politics)* 23: 683–702. https://doi.org/10.13137/1825-5167/32411.

Dove, Edward S., Susan E. Kelly, Federica Lucivero, Mavis Machirori, Sandi Dheensa, and Barbara Prainsack. 2017. Beyond individualism: Is there a place for relational autonomy in clinical practice and research? *Clinical Ethics* 12: 150–165. https://doi.org/10.1177/1477750917704156.

Emanuel, Ezekiel J., and Linda L. Emanuel. 1992. Four models of the physician–patient relationship. *JAMA* 267 (16): 2221–2226. https://doi.org/10.1001/jama.1992.03480160079038.

Entralgo, Lain. 1969. *Doctor and patient*. New York: McGraw-Hill Book/World University Library.

Esposito, Roberto. 2015. *Persons and things: From the body's point of view*. Cambridge: Polity.

Faden, Ruth R., and Tom L. Beauchamp. 1986. *A history of theory of informed consent*. New York: Oxford University Press.

Ferretti, Giovanni. 1996. *La filosofia di Levinas: Alterità e trascendenza*. Torino: Rosenberg & Sellers.

Fried, Charles. 1974. *Medical experimentation: Personal integrity and social policy*. New York: American Elsevier Publishing.

Gilligan, Carol. 2003. *In a different voice. Psychological theory and women's development*. Cambridge: Harvard University Press.

Hardwing, John. 1990. What about the family? *The Hastings Center Report* 20: 5–10. https://doi.org/10.2307/3562603.

Held, Virginia. 1987. Non-contractual society: A feminist view. *Canadian Journal of Philosophy* 17: 111–137. https://doi.org/10.1080/00455091.1987.10715932.

Jennings, Robert. 2016. Reconceptualizing autonomy: A relational turn in bioethics. *The Hastings Center Report* 46 (3): 11–16. https://doi.org/10.1002/hast.544.

Kittay, Eva Feder. 1999. *Love's labor. Essays on women, equality, and dependency*. London: Routledge.
Lauritzen, Paul. 1989. A feminist ethic and the new romanticism – Mothering as a model of moral relations. *Hypatia* 4: 29–44. https://doi.org/10.1111/j.1527-2001.1989.tb00571.x.
Levinas, Emmanuel. 2006. *Humanism of the other*. Champaign: University of Illinois Press.
Levinas, E. 2007. *Totality and infinity. An essay on exteriority*. Pittsburgh: Duquesne University Press.
———. 2011. *Otherwise than being or beyond essence*. Pittsburgh: Duquesne University Press.
Mackenzie, Catriona, and Natalie Stoljar. 2000. *Relational autonomy. Feminist perspectives on autonomy, agency and social self*. Oxford: Oxford University Press.
Mahowald, Mary. 1987. Sex-role stereotypes in medicine. *Hypatia* 2 (2): 21–38. https://doi.org/10.1111/j.1527-2001.1987.tb01063.x.
Manson, Neil C., and Onora O'Neil. 2007. *Rethinking informed consent in bioethics*. Cambridge: Cambridge University Press.
Noddings, Nel. 2013. *Caring. A relational approach to ethics and moral education*. Berkeley: University of California Press.
Nussbaum, Martha. 2022. *Upheavals of thought: The intelligence of emotions*. Cambridge: Cambridge University Press.
Palumbo, Rocco. 2017. *The bright side and the dark side of patient empowerment. Co-creation and co-destruction of value in the healthcare environment*. Cham: Springer.
Pulcini, Elena. 2009. *Care of the world: Fear, responsibility and justice in the global age*. Dordrecht: Springer.
Ricoeur, Paul. 1991. Love and justice. In *Radical pluralism and truth: David Tracy and the hermeneutics of religion*, ed. Jeanrond Werner and Rike Jennifer, 187–202. New York: Crossroad.
Rosato, Jennifer. 2012. Woman as vulnerable self: The trope of maternity in Levinas's *otherwise than being*. *Hypatia* 2: 348–365. https://doi.org/10.1111/j.1527-2001.2011.01189.x.
Ruddick, Sarah. 1983. Thinking about mothering – And putting maternal thinking to use. *Women's Studies Quarterly* 11: 4–7.
———. 1990. *Maternal thinking: Towards a politics of peace*. London: Women's Press.
Sherwin, S. 2008. Whither bioethics? How feminism can help reorient bioethics. *IJFAB: International Journal of Feminist Approaches to Bioethics* 1 (1): 7–27. https://doi.org/10.3138/ijfab.1.1.7.
Siegler, Mark. 1984. Should age be a criterion in health care? *The Hastings Center Report* 14: 24–27. https://doi.org/10.2307/3561093.
Siegler, Mark, and James F. Childress. 1984. Metaphors and models of doctor–patient relationships: Their implications for autonomy. *Theoretical Medicine* 5: 17–30. https://doi.org/10.1007/BF00489243.
Sullivan, Laura S. 2016. Medical maternalism: Beyond paternalism and antipaternalism. *Journal of Medical Ethics* 42: 439–444. https://doi.org/10.1136/medethics-2015-103095.
Tang, T.S., M.M. Funnell, M.B. Brown, and J.E. Kurlander. 2010. Self-management support in "real-world" settings: An empowerment-based intervention. *Patient Education and Counseling* 79: 178–184. https://doi.org/10.1016/j.pec.2009.09.029.
Tauber, Alfred. 2005. *Patient autonomy and the ethics of responsibility*. London: MIT Press.
Veatch, Robert. 1972. Models for ethical medicine in a revolutionary age. *Hastings Center Report* 2: 5–7. https://doi.org/10.2307/3560825.
Verkek, Marian. 2001. The care perspective and autonomy. *Medicine, Health Care and Philosophy* 4: 289–294.
Whitbeck, C. 1984. The maternal instinct. In *Mothering: Essays in feminist theory*, ed. Joyce Treblicot, 185–191. Totowa: Roman and Allanheld.
Wolpe, Paul R. 1998. The triumph of autonomy in American bioethics: A sociological view. In *Bioethics and society: Sociological investigations and the enterprise of bioethics*, ed. Raymond De Vries and Janardan Subedi, 38–59. Englewood Cliffs: Prentice Hall.

Chapter 3
A Decolonial Ethics-Aesthetics of Care

María Grace Salamanca González

Abstract The current planetary context reveals an urgency to create a philosophy committed to our world. In this context, and building upon Latin-American intellectual history, this chapter develops a proposition that is both theoretical and practical: the aesthetics of care. This proposition is built upon the imaginary dimension of our lives and, therefore, of the crises, particularly those raised by the Anthropocene epoch. This proposition involves practicing the situated thought in different registers, which will be developed throughout the text. In summary, it involves *sentipensar* the crisis, that is, not only thinking about it but feeling it at the same time, treating both acts with equal importance. The chapter is organized into two parts. On the one hand, a decolonial reading on the ethics of care is presented, articulated in three axes: the human condition understood as vulnerable, the dynamics of attention and inattentiveness typical of the modern world-system, and the contextualization of decision-making. On the other hand, the proposition of an aesthetics of care is described along two axes: firstly, the understanding of *aesthetics* from a decolonial perspective, and finally, the development of my own practice of the aesthetics of care through theater.

Keywords Ethics of care · Aesthetics of care · Decolonial ethics · Latin-American ethics · Anthropocene

M. G. Salamanca González (✉)
Departamento de Historia y Filosofía de la Medicina, Facultad de Medicina, Universidad Nacional Autónoma de México, Ciudad de México, México

J. Braga (ed.), *Ethics of Care*, Advancing Global Bioethics 20,
https://doi.org/10.1007/978-3-031-73608-7_3

1 Introduction

Fragility, vulnerability, and uncertainty could be postulated as characteristics of our time. We have faced a planetary health crisis; unfortunately, it is not the only current life-threatening problem. We are confronted with an overlay of emergencies that compromise the earth's habitability.

The reflection on the dangers and possibilities for life on the planet is *becoming an event* under the concept of "Anthropocene", which can be defined as the period in which life on earth is threatened for the planet as a whole, due to beliefs, representations, and human decisions made by the instrumental rationality of modernity, in its ambition for domination, particularly of what was designated as the "natural" world (Salamanca González 2023).

There is an imaginary dimension of the crisis that involves how we think about them, how we conceptualize them, but also how we feel them, and how we (if even there is "we") perceive them (Glissant 1990). If we recognize the importance of the images and representations used to describe the problems and their possible solutions, we may wonder if there is an imaginary dimension to the current crisis, if the fact of continuing to think with modern concepts and values prevents us from imagining solutions outside the modern world-system in a way that the existing system is perceived as the reality, imprisoning all possibilities to exit. Maybe the cage is open, but it is as if it was still closed (Santos De Sousa 2018).

One of the objectives of this text is to stop bouncing within the walls of our modern cage. I am not going to describe nor criticize modernity. My goal here is to multiply the options (Mignolo and Gómez Moreno 2012), which implies narrating what we have built to face our crisis, in our circumstances, with the hope of encouraging the creation of new alternatives.

> For us, the decolonial proposal is presented as an option and not as an imposition that seeks to take the place of the adversary. On the contrary, the decolonial proposal wants to show that an adversary is always an option, not the only option[1] (Mignolo and Gómez Moreno 2012, 11).

The following pages aim to open possibilities for diversities, in the famous Zapatista's formulation "To make worlds in which many the worlds fit" (Comité Clandestino Revolucionario Indígena-Comandancia General del Ejército Zapatista de Liberación Nacional 1996). To achieve that, I inscribe this proposition in the epistemic perspective Pascale Molinier has named as a "strategic optimism", which starts by acknowledging how, while doing research, one possibility is to study the crises to better understand them, but there is also the option to study the ways to get around them. Particularly, she proposes to study how to pay attention, and care, are paths away from violence (Molinier 2018).

[1] Translated from Spanish by the author: "Para nosotros, la propuesta decolonial se presenta como una opción y no como una imposición que pretende ocupar el lugar del adversario. Todo lo contrario, la propuesta decolonial quiere mostrar que el adversario es siempre una opción, no la única opción".

The proposition presented in this paper consists of a decolonial reading of the ethics of care. Building up on Pascale Molinier's argument, I am going to explore how a different distribution of attention and inattention may lead to a different value scale which may represent an alternative to keep our world habitable.

Finally, this proposal is located in Latin-American intellectual history, which means as researchers, our task is perceived to not only describe our worlds, but to think *from-with-for-between* our realities, and to fully do it, it is important to situate our thought.

To situate our thought is not only a way of showing the limits and scopes of our perspective, but it is also offering the reader some clues to comprehend an intellectual history that may not be her own, a way of thinking which may not correspond to hers, to address realities that may not be known or problems that may remain hidden. By practicing situated thought, we aim not only to share ideas but to also share the hermeneutic tools to make an intercultural dialogue possible, especially when the writer is not part of the dominant culture.

To describe my place of enunciation I would like to say that I live between what would be considered by a certain logic as contradictory borders. I live in a borderland (Anzaldúa 1987) at the borders of modernity: I identify myself as a woman, I'm Mexican (from the province), I'm a mother, neurodiverse, perceived most of the times as white, I have a French Ph.D., I'm not rich, I'm educated in philosophy and social sciences, I'm also a theater actress, and finally, particularly important for this text: I do not identify myself either as modern, postmodern, or alter-modern.

I know this may seem to the reader, at first glance, completely off-topic. My goal here seems to be to do philosophy, to think unattached from myself, to take some distance, and to leave my personal life out of these lines. But not only am I not able to do it (because I am not modern), but I do not want to do it. Therefore, I am going to talk about myself, I am going to talk from my point of view, shaped by my nationality, my culture, and my history.

To summarize, this is an attempt to face the critical urgencies of our time. This proposal implies to *sentipensar*[2] the crisis, that is not only to think about them but to feel them, at the same time and treating both acts with the same status. That is why this paper does not only address an ethic of care, but also an aesthetic. Because once decentered from modernity we realize the emotional dimension of life is crucial to the decision-making processes, and from a decolonial point of view, it is not only important to state it but to practice it.

To develop this last dimension, and to try a cultural translation of the meaning behind the *sentipensar* I talk of ethics-aesthetics. In this context, the *aesthetics of care* are not only a concept but a practice. In my experience, it is a concept-practice developed through theater.

[2] Neologism attributed to the fishers of Colombia, and currently used in all Latin America. It combines in one word the verbs to feel and to think: to *feelthink*.

To present what *sentipensar* the crisis means I will divide this paper into three parts: first I will present a decolonial reading of the ethics of care, then I will develop the aesthetics of care, and finish with their articulation.

Finally, I would like to insist on how this proposal is only that: an option. Of course, it is not the best, a better, universal, or the only option. What I would like to highlight is how a current world-modern system is also an option, not the best, nor a universal option, and even if the modern world is real, certainly, it is not the only reality.

2 A Decolonial Ethics of Care

The ethics of care is a moral theory that values the contextualization of decisions, the emotional context shaping decision-making processes, the diversity of moral values, and the interdependence of life in general and human life in particular (Tronto 1993; Gilligan 2008). According to this theory, care will be understood as both a disposition and a practice (Held 2006).

Some of the developments in the ethics of care allow us to imagine non-colonial ethics, based on the possibilities for moral diversities once accepting there is not only one universal moral voice but many (Gilligan 2008). The question of moral diversity is central to a theory developed from-for Latin America, particularly in Mexico because we live together with nations preexisting colonization.

In this context, in the objective of modeling non-colonial ethics for diversities, it becomes important to get away from the civilizing and universalist readings of morality, which exist also in some conceptualizations of the ethics of care. Thus, a decolonial reading on the ethics of care has these goals in mind: develop non-colonial ethics and create ethics compatible with an intercultural project (Salamanca González 2020).

The decolonial reading on the ethics of care has been developed by me and by members of the research program Social Actors of the Medicinal Flora in Mexico, part of the National Institute of Anthropology and History (Salamanca González and Hersch Martínez 2019; Hersch Martínez and Salamanca González 2021). We elaborated this proposal by knitting the ethics of care with the sociocultural epidemiology around three axes: relational anthropology, the attention and inattention dynamics structured by the modern world-system, and the importance of the contextualization of decisions (Salamanca González 2021).

To address the objective of this paper I will elaborate on the decolonial ethics of care in three stages: first, the importance and decolonial meaning of the particular turn of morality; second, a conceptualization on the moral subject; and finally, our distinction between attention, inattention, and the neologism in English *desattention*.

2.1 The Particular Turn of Morality

The ethics of care could be understood as a particular turn of morality (Brugère 2011), moral judgments will be validated by their adjustment to circumstances, the sociocultural context, and the emotional environment. This morality could be resumed in the phrase "it depends" which seems to characterize a contextual *different moral voice* (Gilligan 2008).

The ethics of care is a moral theory evaluating contextualized judgments and contextualized opinions. The moral value changes from searching for a general (supposedly universal) justice to searching for the adjustment to circumstances. To summarize this idea:

> Reducing ethics to the individual choices of a person, which in turn are supposed to emanate from a deep reflection, whose foundations would be philosophical and psychological, and influenced by religious or secular values in a quest for universality, simply disregard the social processes of moral life. These processes show the extent to which the individual is caught up in economic, cultural, family, friendship, and professional activities which forcefully define his or her moral horizon in a way of which he or she is probably only partly aware. Finally, ethics, if we conceive it as a model of moral reasoning that claims to be the champion of abstract reflection and the rational choice of autonomous individuals in search of objective moral criteria, runs the risk of being unsuitable to the framework of human experiences which take place almost always in uncertain circumstances and quite specific contexts (Kleinman 2013, 401).[3]

A decolonial ethics of care recognizes the importance of contextual adjustments for our moral reflections, but it starts applying that same reasoning to the theorizing activity itself. It begins by affirming how it is impossible to speak from God's point of view, and the voice –and person– theorizing is not an overhanging reality, but it is tied to a specific context. This leads the author to theorize some subjects over others, to consider some problems, to analyze with a certain theoretical framework, to be familiar with certain authors (and ignore some others), and to consider certain theories as valid interlocutors while ignoring others.

The decolonial theory states that all ideas, thoughts, and therefore theories, have a place of enunciation. Santiago Castro Gómez names *the hybris of point zero* to the power device that imposes some voices as not affected by their circumstances as if some authors could think and theorize independently about their experiences. As if

[3] Translated from French by the author: "Réduire l'éthique aux choix individuels d'une personne, qui sont à leur tour censés émaner d'une réflexion profonde, dont les soubassements seraient philosophiques et psychologiques, et influencés par des valeurs religieuses ou laïques dans une quête d'universalité, ne rend tout simplement pas compte des processus sociaux de la vie morale. Ces processus montrent à quel point l'individu est pris dans des activités économiques, culturelles, familiales, amicales et professionnelles qui définissent avec force son horizon moral d'une manière dont il ou elle n'est probablement qu'en partie conscient(e). Finalement, l'éthique, si on la conçoit comme un modèle de raisonnement moral qui se fait fort d'être le champion de la réflexion abstraite et du choix rationnel d'individus autonomes en quête de critères moraux objectifs, court le risque d'être inadaptée au cadre d'expériences humaines qui se déroulent dans des circonstances presque toujours incertaines et dans des contextes tout à fait spécifiques".

problems unknown by authors could be adequately addressed by them as if theories unknown by authors could be reducible to the hegemonic concepts those authors have read and used. This dimension is conceptualized in decolonial theory as *the coloniality of knowledge* (Castro Gómez 2000).

Coloniality differs from colonialism in the sense that colonialism refers to the military, economic, social, and political occupation of land, bodies, and living and non-living things. While coloniality refers to the hierarchy imposed by colonialism, in which the colonizer and what he represents, knows, believes, and likes are morally (and universally) better than those of the colonized. It becomes important to highlight that this superiority is built on the rejection of knowing the other: it is completely ignored how the colonized represents herself, how she identifies herself, what she believes, and what she knows or likes but it is assumed as naturally inferior to the colonizer. Thus, coloniality is a technology of power based on an *alleged* knowledge of the other (Castro Gómez 2000).

In the context of knowledge production, the coloniality of knowledge is sustained by the separation between the author, her interest, her education, her nationality, her gender, her object of study, her funding, and the contents of her research. To fight the coloniality of knowledge, it is proposed to situate our thinking: to describe who we are, what our interests are, our intellectual history, and the interlocutors we are considering (Corona Berkin 2020).

To situate our thinking, we may reflect on these questions: who is speaking and from what point of view? Because we all have different identities, I could be writing these lines as a philosopher or as a theater actress, for example. And in function of that answer, my words could –and maybe should– be interpreted differently.

Another dimension to question is if the voice is used in a particular register (in the first person singular) or if the author is trying to represent others. If it is the latter, we may ask if the author has consent to represent a social group or if she is speaking on their behalf without asking them. It is central to remember that one of the first strategies of coloniality is to believe that some cannot speak for themselves, who cannot even understand themselves, who are unable to produce explanations, even on their behalf. Therefore, a decolonial theory cannot speak in the name of others (Corona Berkin 2020).

Another question to consider is the direction of the reflection: what are the topics susceptible to be considered as "ethical issues", and what is the framework used to approach them? Aimé Césaire has argued how some genocides are recognized and constitute a moral scandal, but some others continue to be denied or underestimated (Césaire 2004). Given that it is impossible to consider everything, it becomes central to delimit what we are talking about, and from what perspective (Salamanca González 2020).

Finally, to situate our discourse, it may be important to question whom we are addressing and what are we trying to accomplish by talking to that specific audience. This becomes an important question if we consider how many times authors analyze moral situations outside their sociocultural context to be discussed without the directly involved. Coloniality of knowledge makes it morally acceptable to discuss how would it be fair to distribute humanitarian help in the African continent, and that course is systematically being taught in first-world countries. We may ask ourselves about the justification of that audience.

In this section, I have described the importance of decolonial ethics to situate the reflection. The objectivation –and universalization– of morality is seen here as a colonial practice, as an exercise, and as a strategy, for domination. Mexican popular knowledge has a saying to summarize this problem it is as "throwing a rock and hiding the hand". For decolonial theory, the fact of speaking without clearly delimiting the scope and the limits of the discourse's author is throwing the rock and trying to hide the hand. In other words, it implies denying or reducing the responsibility of the author behind the theory.

From our perspective, the particular turn of morality is not only reduced to the content of the moral judgments and the conditions for their validations, but we are proposing to enlarge the scope to reach the production of ethical theory. This proposition implies "naming the responsibility" and argues that accountability starts by clarifying who we are, our interests, our problems, our interlocutors, and our audiences, and of course, it implies being accountable for our concepts.

2.2 *On the Moral Subject*

One of the fundamental critiques of the ethics of care is the refusal to accept the human condition as reducible to the ideas and characteristics of the most dominant individuals in societies. First, Carol Gilligan questions the gender assumption, the fact of conceiving the male sociocultural education as the universal moral standard:

> It is difficult to say 'different' without implying 'better' or 'worse', as we tend to construct a single scale of measurement and as this scale has generally been developed and standardized from data interpreted by men and obtained as a result of surveys, whose participants were mostly or exclusively male, psychologists "have been predisposed to view male behavior as the 'norm' and female behavior as a kind of deviance". Therefore, when women do not conform to the standards set by psychologists, the usual conclusion has been that something is wrong with them (Gilligan 2008, 32).[4]

This is the first "different voice" founding the ethics of care. Afterward, other authors have deeply developed this critique: Joan Tronto has successfully shown how the present moral boundaries sustain the contemporary political institution and its power relations (Tronto 1993) in the field of philosophy. Virginia Held has

[4]Translated from French by the author: "il est difficile de dire « différent » sans impliquer « mieux » ou « pire », comme nous avons tendance à construire une seule échelle de mesure et comme cette échelle a été généralement élaborée et standardisée à partir de données interprétées par des hom mes et obtenues à la suite d'enquêtes, dont les participants étaient principalement ou exclusivement des hommes, les psychologues « ont été prédisposés à considérer le comportement masculin comme la « norme » et le comportement féminin comme une sorte de déviance ». Par conséquent, lorsque les femmes ne se conforment pas aux standards établis par les psychologues, la conclusion habituelle a été que quelque chose n'allait pas chez ces dernières".

constructed a strong critique on the supposed subject's universality shaping moral theories (Held 2006).

> Kohlberg's theory only tells the story of moral development from a standpoint of those who have remained on top throughout the entire process. Those who have stumbled need to forget their experiences and become like the successful ones. Those who try to tell the story of racism, sexism, ethnocentrism, and so forth, are misunderstood or derided for their inability to get beyond these sticking points (Tronto 1993, 73).

The first condition to be listened to, to be taken seriously, is to be rational, and to be rational is reduced to being modern, to be a part of (or to imitate) the dominant-modern model. It is as if the entrance condition to a public discussion on ethics was to be a modern man (Molinier 2018). To be taken seriously, or to "be realistic", all diversities will need to camouflage to the hegemonic model, loss the difference, and drop the accent.

A decolonial ethics of care is built on human diversity and strives to enhance it. Decolonial ethics starts by affirming how our possible experiences of the world are vaster than the modern's explanation of them (Santos 2009). Decolonial ethics looks back at history and can spot how the modern human condition was modeled to contrast what was considered not modern: what and those who were historically built as savages, as irrationals, as needing to be civilized by the violence of others (Mignolo 2003, 2005).

There is a colonial matrix in the assumption of one univocal human condition because all of those not considered, or all of those not fulfilling the imposed characteristics, will be conceptualized, and therefore treated, as inferiors. The possible categories to inferiorize others could be their gender, age, skin color, socioeconomic class, relation with work (having a job, and even a certain type of job), inhabiting in specific ways (houses, cities), transporting in certain ways, eating certain things in specific manners, speaking certain languages in precise ways, being psychologically and neurologically considered as part of normality, having access to running water, to electricity, to internet…

In other words, it remains the generalized assumption of the western human model as being the human condition. This assumption works in a colonial way inferiorizing diversities. This statement leads to affirming that western values are universal and leads to the belief that the subalternized others wish to resemble their colonizers as much as possible.

This last idea is central to the decolonial theory, the coloniality of the being is not only reduced to inferiorize the non-modern, but in the same movement, it raises the modern model as the norm (Castro Gómez 2005). This is why some authors claim that modernity and coloniality are two sides of the same movement because the superiority of modernity was built by differentiating it from what was not and those that were not modern (Quijano 1991). That is to say, the historically-built superiority of modernity was constructed on the historically-built inferiority of others, inflicting the non-moderns the *colonial wound:* labeling their differences as inferior and irreparable (Anzaldúa 1987; Mignolo 2003, 2005).

The critique on the univocity of the moral subject is nodal to decolonial ethics because only from this perspective we are also fully humans. This critique is essential to the emancipation from the coloniality of being, to embrace our differences. The acceptance of our reasons, our logics, our feelings, our rationalities, our priorities, our problems, and our voices, constitute a way of becoming responsible for them.

2.3 Attention, Inattention and Desattention

The concept of attention has been central to the development of the ethics of care. As stated above, we define care both as a practice, as the actions we realize to answer to detected needs, and also as a disposition. The ethics of care have argued how we only respond to perceived needs. Therefore, one of the most powerful strategies not to care is not to perceive (Tronto 1993; Molinier 2013).

Under the concept of *privileged irresponsibility,* Joan Tronto has developed the strategies the dominants use to maintain the current moral boundaries: the alleged universality of morality, and the human model as independent and autonomous (Tronto 1993). Pascale Molinier has built upon that argument to describe how the opposite of care many times is not directly to harm but to ignore or to be indifferent towards others, especially if those others are subalternized (Molinier 2013).

> The indifference of the privileged is a psychological posture that is based on a complex device—ideological but also of segregation—which prevents the privileged from thinking about dimensions of reality that could embarrass them or prevent them from enjoying their privileges or harm their sense of justice. It is of the greatest comfort to ignore who serves us and what it costs her to do so. Ignorance, Mills insists on the cognitive dimension. In indifference, Tronto emphasizes the affective dimension. *I don't care*, in French, can literally be translated as "Je m'en fiche". Anything unpleasant or disagreeable may well happen to the other, which is indifferent to me. It doesn't affect me. And then, who are you telling me about? This apathy is not only based on a set of errors and lack of thinking, but it also involves a lack of perception. Perception is an active process that involves volition. Thus, an object a person can be in the field of vision, but not being perceived (Molinier 2013, 65).[5]

[5]Translated from French by the author: "L'indifférence des privilégiés est une posture psychologique qui repose sur un dispositif complexe –idéologique mais aussi de ségrégation– qui empêche les privilégiés de penser des dimensions de la réalité qui pourraient venir les embarrasser ou les empêcher de jouir de leurs privilèges ou mettre à mal leur sens de la justice. Il est du plus grand confort d'ignorer ce qui nous sert et ce qu'il lui en coûte de le faire. Ignorance, Mills insiste sur la dimension cognitive. Indifférence, Tronto met l'accent sur la dimension affective. « I don't care », en français, cela peut se traduire littéralement par « Je m'en fiche ». Il peut bien arriver à l'autre n'importe quoi de déplaisant ou de désagréable, cela m'est indifférent. Cela ne me touche pas. Et puis, de quel autre me parles-tu? Cette apathie ne repose pas seulement sur un ensemble d'erreurs et de manque à penser, elle implique aussi un défaut de perception. La perception est un processus actif qui implique la volition. Ainsi un objet, une personne peuvent être dans le champ de vision, mais ne pas être perçus".

Depending on our sociocultural context we may or may not be educated to perceive needs. It is impossible to pay attention to all details of our realities, so we are educated to ignore people, genders, objects, countries, and beings according to the dominant values. This social construction of inattentiveness, ignorance, and indifference seems to be able to justify the privileged irresponsibility. It seems responsibility can only be assigned to actions and not to absences or omissions. How can we call someone responsible for something invisible to her?

According to the theory of the Epistemologies of the South (Santos 2009) this practice could be interpreted as the *production of absences*, by this concept Santos argues that the ignored aspects of reality are actively produced as non-existing, non-important, and off-topic. To clarify this concept, I prefer the word in Spanish "desatención" which differs from inattentiveness and indifference. The neologism in English *desattention* implies an aspect of reality that we may be able to accept, we may even accept its importance, but we are going to permit ourselves to ignore it (Salamanca González 2021). Normally, we will argue there are other more important, more interesting, more urgent, topics to discuss.

Under the colonial context, once colonialism was over, when the subalterns arrive to raise questions that may embarrass or confront the privileges of the dominants, the strategy has been not necessarily to ignore them, but to disregard the arguments as off-topic. In other words, to re-establish the instituted attention's hierarchy. When the subaltern's priorities do not follow the agenda of the privileged, the strategy has been *desattention*: to acknowledge the problems raised followed by a call back to order, back to the topics of the agenda which are –and should remain– somewhere else. *Desattentions* as power strategies imply acknowledging the problems to, immediately after, look somewhere else, that is to reestablish the dominant's distribution of attention.

3 A Decolonial Aesthetics of Care

Perhaps the main characteristic of the ethics of care, particularly while not modern, is the recognition of the value of the emotional circumstances shaping the decision-making processes. The different moral voices described by Carol Gilligan valued the adjustment to the circumstances, not only to the material and symbolic dimensions of realities but an emotional dimension was pointed out.

> These practices [of care], in fact, are addressed to living beings understood from their dependence and their vulnerability, in such a way that everyone interested in these practices, as actor or spectator, is called to measure their dependency and vulnerability. As a result, a theory that takes the importance of care seriously cannot disregard the psychology of emotions (Molinier in Laugier 2012, 88).[6]

[6]Translated from French by the author: "Ces pratiques, en effet, s'adressent à des êtres vivants compris à partir de leur dépendance et de leur vulnérabilité, de telle sorte que chacun qui s'intéresse à ces pratiques, comme acteur ou spectateur, est convoqué à prendre la mesure de sa propre dépen-

From a decolonial standpoint, this reflection is central because it allows us to integrate the corporal and sentimental dimensions as important parts of our lives. To link this argument from the ethics of care to the decolonial theory I thought of the Latin-American concept of *sentipensar*. It involves thinking and feeling as two stages of the same act. From my perspective, this was the adjustment Carol Gilligan and Pascale Molinier were describing, which involved not only responding to circumstances in an intellectual register but on an emotional one.

However, to keep naming the proposition as "an ethics of care" was understood as if this reflection was only that: a reasoning susceptible to be thought, with internal logic, from which some theoretical –rational– conclusions were inferred. Eventually, I started to relate the ethics of care propositions on attention, *desattention*, listening, and adjustment to my theater practices. And I realized it was more precise to name this practical reflection "an aesthetics of care" when we are not only describing cases or realities but when we are hoping to have an incidence on sensibilities and how they are shaped.

To fully understand how the aesthetics of care were born it is necessary to go back to the beginning, to recall how I, educated in Latin America and its critical theory, have always practiced research engaged in its sociocultural context. Eventually, I was asked by the communities to work through theater, and while I was learning about the ethics of care.[7]

I realized how some of the exercises, the emotional dispositions, the social games, the use of the body, and the atmosphere created by the theater workshops could be interpreted as an aesthetic of care, which is not just trying to understand or study how decisions were made, but is trying to shape how we feel, what and how we perceive, how we interpret, and how a situation can be represented and interpreted in different ways (Salamanca González 2021).

> [the aesthetics of care are a] training of the sensibility oriented towards attention, the concern for others, towards listening, towards the effort to relate to others. It is an aesthetic because we are going to study its sensitive forms, as a way of interpreting our lives, our behaviors, and our ways of interacting with the otherness of the world (not necessarily only humans) (Salamanca González 2021, 305).[8]

In this context, and following the proposition of decolonial aesthetics, the meaning of "aesthetics" does not reduce to art, nor to beauty or what could be appreciated as sublime. The aesthetics from a decolonial perspective will maintain the Greek meaning of senses and emotions (Mignolo in Mignolo and Gómez Moreno 2012). To clarify our understanding of this concept we could relate "aesthetics" with the

---

dance et de sa propre vulnérabilité. Il en résulte qu'une théorie qui prend au sérieux l'importance du care ne peut faire abstraction d'une psychologie des émotions".

[7] In the first chapter of my Ph.D. thesis, I develop how the relationship with the communities was established, and how the articulation theory-practice was built. Cf. (Salamanca González 2021).

[8] Translated from French by the author: "formation sensible tournée vers l'attention, le souci des autres, vers l'écoute, vers l'effort de se mettre en relation avec autrui. Il s'agit d'une esthétique en ce que nous allons en étudier les formes sensibles, ainsi que la manière d'interpréter sa vie et ses conduites et d'interagir face à l'altérité du monde (pas nécessairement seulement humaine)".

word "senses" in its two meanings: on the one hand, the perception as in the five senses, and on the other the social construction of sense.

> [...] decolonize aesthetics to liberate the aesthesis. To liberate the aesthesis means performing two simultaneous movements; on the one hand, to deprogram modern and postmodern philosophy that takes refuge in the regionality of the western concept of aesthetics and, on the other, to deprogram aesthetics that reduces aesthesis to the sphere of art: aesthesis, feeling, the senses affect us at all times of existence. Aesthesis is not reduced to the beautiful and the sublime, but to the processes and events by which certain actors and institutions manipulate the senses; that is, the responses of bodies to stimuli (Mignolo 2015, 435).[9]

In the practice of the aesthetics of care, we will experiment through theater exercises on how our perceptions are constructed, how we feel, what we do not perceive nor feel, the patterns we use to pay attention and ignore, and how we listen (or not). We will explore our sensibilities through theater, and collectively discuss them, not just to gain understanding, but to change what we consider important to change.

The practice of the aesthetics of care builds its justification on the expansion of our possibilities to feel, represent and express. As well as in the opening of spaces to collectively reflect on our perceived problems and the search through the body and always with the group of new ways to interpret and to solve them. And finally, the aesthetics of care gain their meaning on the opening of new possibilities for dialogues which are *sentipensados*, that is that they are including not only reasons and arguments, but our feelings (Salamanca González 2023).

The practice of the aesthetics of care demands an openness to uncertainty: not having a preestablished plan for the workshops or shows to privilege the cocreation with the participants. In the practice of the aesthetics of care, there is nothing preceding the encounter. The first step will be to sensitize ourselves to be receptive to others, being open, available, and attentive with all our bodies and gestures (Salamanca González 2023).

Above this openness, the non-colonial listening and interpreting will also be constructed, because for us to listen and interpret alterity on our own terms and to do so, we need to work on our ability to suspend pre-judgments:

> To suspend all judgment of reality to be able to listen to statements or facts of language, themselves inseparable from specific intensities (or emotions), that is to say not only to allow oneself to be affected but to also simultaneously take words seriously, to *enter into the forms of life* (Molinier 2018, 81).[10]

[9] Translated from Spanish by the author: "descolonizar la estética para liberar la aesthesis. Liberar la aesthesis significa realizar dos movimientos simultáneos; por una parte, desprogramar la filosofía moderna y posmoderna que se amparan en la regionalidad del concepto occidental de estética y, por la otra, desprogramar la estética que reduce la aesthesis a la esfera del arte: la aesthesis, el sentir, los sentidos nos afectan en todo momento del existir. La aesthesis no se reduce a lo bello y lo sublime, sino a los procesos y acontecimientos por los cuales determinados actores e instituciones manipulan los sentidos; es decir, las respuestas de los cuerpos a los estímulos".

[10] Translated from French by the author: "Suspendre tout jugement de réalité pour être en mesure d'écouter des dires ou des faits de langage, eux-mêmes indissociables d'intensités spécifiques (ou émotions), c'est-à-dire non seulement se laisser affecter, mais simultanément prendre les dires au sérieux, pour entrer dans des formes de vie".

From this perspective, we cannot assume a "neutral listening" was the universal hermeneutic condition. On the contrary, we argue there is a communicational vulnerability, which implies our words, interpretations, reactions, and gestures could bless the other's discourse. Furthermore, we argue symbolic wounds could be irreparable. So, when we care about words, it means to also care about how we listen, which leads us to propose a politics of listening: to suspend judgments and to take the other seriously in her difference, in her alterity (Molinier 2013).

> We train ourselves in this work of learning to live with the opacity implied by otherness, in the impossibility of complete epistemic access to the other's psyche to understand her story in her terms. We are ourselves; we have the possibilities offered by our body, our tone, and our gestures. [...] There is indeed a spirit of domination and control in the ambition to vampirize the other, to break into her to steal from her, to understand everything about her, as if it was possible, already for herself, to have transparent access to her psyche, motives, and actions. Even more domination in the appropriation's desire! To argue that my interpretation of her is actually her. Not only the ambition to assimilate otherness to me, but to believe that my possibilities (psychic and physical) allow me a "transparent", "neutral", and "mirror" type representation of the other. Colonial epistemologies (Salamanca González 2021, 324).[11]

In the theater practice of the aesthetics of care, we work with the real stories told by the participants, and we train ourselves in political listening, one in which it is not as important to find an alleged univocal truth but to show how diverse interpretations of one single story are possible.

To summarize, the aesthetics of care are firstly a practical alternative to try different possibilities of perceiving, feeling, expressing, representing, and responding to particular stories faced by the practitioners. Additionally, the practice of the aesthetics of care allows us to explore the concepts of the ethics of care, which opens roads to specify, develop or add new concepts. In other words, the decolonial practice of the aesthetics of care is a way of *sentipensar* our shared realities.

[11] Translated from French by the author: "Nous nous entrainons dans ce travail à apprendre à vivre avec l'opacité qu'implique l'altérité, dans l'impossibilité d'un accès épistémique complet à la psyché de l'autre et à son histoire dans ses termes propres. Nous sommes nous-mêmes, nous avons les possibilités que nous offrent notre corps, notre ton, nos gestes. [...] Il y a en effet un esprit de domination et de contrôle dans l'ambition de vampiriser l'autre, de faire effraction en lui pour le voler, tout comprendre de lui, comme s'il était possible, déjà pour lui-même, d'avoir un accès transparent à sa propre psyché, à ses motivations et actions. Domination encore dans ce désir de nous l'approprier! De soutenir que mon interprétation de lui soit lui. Ce n'est pas seulement l'ambition de m'assimiler l'altérité, mais de croire que mes possibilités (psychiques et physiques) me permettent une représentation « transparente », « neutre », de type « miroir » de l'autre. Épistémologies coloniales".

4 Final Reflections

This research started situated in Latin-American critical theory with the objective of thinking *from-with-for-between* our shared historic worlds. And our context is framed in the Anthropocene, that is to say, our life is threatened by a modern-colonial way of living. The modern way of living embraces individual desires over the planetary boundaries and embraces to keep dreaming without questioning in what conditions the minerals for our high technology were extracted.

A way of living does not only include a material dimension, it is obviously important to reflect on the Anthropocene from the standpoint of the earth sciences, but it has not been sufficient to change individuals' behaviors, or political views to limit the transgression of the planetary boundaries (Steffen et al. 2015). I hypothesize that our modern way of living has an imaginary dimension, a subconscious, which is also colonized (Rolnik 2018), in a way that we keep describing our problems in a modern way, and we keep searching for modern answers to our modern problems.

For instance, in the environmental crisis discussion, the idea that we can think of realities unknown is still assumed. Thinkers located in the global north will just assume the problems and crises they know, study, and live in are either the only ones, the most important, or the most urgent. This proposal argues that to effectively face the crisis, we should situate our statements, which is also proof of intellectual rigor and intellectual honesty.

The imposition of the modern human model as the universal human model, in the context of the Anthropocene, leads to the general assumption we want to "develop" as the "first world": we want to live like them, eat like them, buy like them… This keeps being advertised and assumed even if we acknowledge we do not have the resources, the planetary materials, for everybody to consume and live as the modern model. Therefore, we should revise how our assumption of modernity as the human condition has shaped the current environmental crisis.

Finally, a reflection on Anthropocene from decolonial ethics of care standpoint invites us to reflect not only on what is considered to be central to the crisis, but all those aspects we recognize as being part of the problem, but that we are going to permit ourselves to ignore (Salamanca González 2023). In other words, what are the absences produced by the way we approach the crisis? Is it possible to change the distribution of attention? To what purposes is the current distribution of care helping?

Modern science has not been enough to stop our current crisis and I asked what a decolonial ethics-aesthetics of care could bring to this discussion. I chose not to engage in this discussion to distribute responsibilities and then proceed to label the Epoque as the capitalocene, plantacionocene, or chthulucene. I chose not to focus on the critique of modernity and how its instrumental reason is at the core of the causes of some of the crises because I am not sure if *the master's tools could dismantle his house* (Lorde 2007).

Instead, I decided to speak through my voice, and in line with the intellectual history of my region, I decided to multiply the options. To focus on the existing cracks within the system and explore some of the alternatives. The proposal argues for *sentipensar* our shared worlds.

A decolonial ethics of care is an invitation to enlarge our moral responsibilities and to do so, it is important to decolonize ourselves, that is to say, we need to work for everyone's reappropriation of herself, her own life, her own decisions, actions, and silences. Everyone should feel-know themselves as the expert of herself to enlarge their responsibility in their life. Reclaiming life involves reclaiming one's body, one's sensitivity, one's imagination, one's gestures, one's words, and one's abilities to create one's life and one's future (Salamanca González 2023).

> My ethical proposal founds ethics in aesthetics: I propose to consider perception, sensation, affectability, and sensitive representations as the foundation of our human beings, and as the root of social ties. This ethical voice has the particularity of sensitivity towards others and calls for the development of sensitivity towards others, towards their needs, their expression, their gestures, and their manifestations of life, without renouncing oneself in the expression of one's sensitivity. An aesthetic of relationship. An aesthetic that is liberation because it is not enough to say that care and bonds are important to be in the world, but that it is already important to exercise this attentive sensitive practice, towards oneself and others. It is not enough to recognize the importance of care, it is essential to develop (concretely and not only theoretically) an aesthetic of care, of taking care of others as of oneself (Salamanca González 2021, 337).[12]

Our sensibilities are also the product of our educations, our cultures, of the sociohistorical places we have lived in. We are educated to feel or not to feel, to be indifferent, or to care. The decolonial aesthetics work from the premise that it is possible to decolonize one of our most intimate traits: our sensibility, our way of perceiving and interpreting our worlds.

An aesthetics of care is a sensibilization towards attention, and listening, directed to bodies and their expressions: gestures, silences, and all the means to express which we need to guess, perceive, interpret, suppose, and imagine. Caring is always the realization of aesthetics, and it may as well be one of indifference or domination. Through this proposal, I am trying to focus our ethics of care on the emotional, sensible disposition it implies, not only to document it but to also interfere in it, to decolonize it to gain responsibility through emancipation.

---

[12] Translated from French by the author: "Ma proposition éthique fonde l'éthique dans l'esthétique: je propose de considérer la perception, la sensation, l'affectabilité et la représentation sensible comme le fondement de notre être humain, et comme la racine des liens sociaux. Cette voix éthique a pour particularité la sensibilité envers autrui et en appelle au développement de la sensibilité envers les autres, envers leurs besoins, leur expression, leurs gestes, leurs manifestations de vie, sans renoncer à soi-même dans l'expression de sa propre sensibilité. Une esthétique de la relation. Une esthétique qui est libération parce qu'il ne suffit pas de dire que le care et les liens sont importants pour être au monde, mais qu'il importe déjà d'exercer cette pratique sensible attentionnée, envers soi et envers les autres. Il ne suffit pas de reconnaitre l'importance du care, mais il est indispensable de développer (concrètement et non pas seulement théoriquement) une esthétique du care, du prendre soin, d'eux comme de soi".

References

Anzaldúa, Gloria. 1987. *Borderlands. La Frontera. The New Mestiza*. San Francisco: Aunt Lute Books.

Brugère, Fabienne. 2011. *L'éthique du « care »*. Troisième. Que Sais-Je?. Paris: Presses Universitaires de France.

Castro Gómez, Santiago. 2000. Ciencias sociales, violencia epistémica y el problema de la invención del otro. In *La colonialidad del saber: eurocentrismo y ciencias sociales. Perspectivas latinoamericanas. Sur-Sur*. Buenos Aires: CLACSO, Consejo Latinoamericano de Ciencias Sociales.

———. 2005. *Hybris del punto cero: Ciencia, raza e ilustración en la Nueva Granada (1750–1816)*. Bogotá: Editorial Pontificia Universidad Javeriana.

Césaire, Aimé. 2004. *Discours sur le colonialisme suivi du Discours sur la négritude*. Paris: Éditions Présence Africaine.

Comité Clandestino Revolucionario Indígena-Comandancia General del Ejército Zapatista de Liberación Nacional. 1996. 1996 Cuarta Declaración de la Selva Lacandona. January 1.

Corona Berkin, Sarah. 2020. *Producción Horizontal del Conocimiento*. Alemania: CALAS Maria Sibylla Merian Center, Bielefeld University Press.

Gilligan, Carol. 2008. *Une voix différente pour une éthique du care*. Paris: Flammarion.

Glissant, Édouard. 1990. *Poétique de la Relation*. Paris: Gallimard.

Held, Virginia. 2006. *The ethics of care: Personal, political, and global*. Oxford: Oxford University Press.

Hersch Martínez, Paul, and María Grace Salamanca González. 2021. El cuidado y los procesos de atención-desatención como referentes analíticos y operativos para la salud colectiva. *Rev. Fac. Nac. Salud Pública* 40. https://doi.org/10.17533/udea.rfnsp.e345191.

Kleinman, Arthur. 2013. Repenser la nouvelle bioéthique. In *La question morale*, ed. Didier Fassin, 398–410. Paris: Presses Universitaires de France.

Laugier, Sandra. 2012. *Tous vulnérables ? le "care", les animaux et l'environnement*, Petite Bibliothèque Payot 840. Paris: Editions Payot & Rivages.

Lorde, Audre. 2007. The master's tools will never dismantle the master's house. In *Sister outsider: Essays and speeches*, 110–114. Berkeley: Crossing Press.

Mignolo, Walter D. 2003. *Historias locales diseños globales : colonialidad, conocimientos subalternos y pensamiento fronterizo*. Madrid: Akal.

———. 2005. *The idea of Latin America*. Blackwell.

———. 2015. *Habitar la frontera. Sentir y pensar la descolonialidad (Antología, 1999–2004)*. Barcelona: CIDOB y UACI.

Mignolo, Walter D., and Pedro Pablo Gómez Moreno, eds. 2012. *Estéticas y opción decolonial*. Bogotá: Universidad Distrital Francisco José de Caldas.

Molinier, Pascale. 2013. *Le travail du care*. Le Genre Du Monde. Paris: La Dispute/Snédit.

———. 2018. *Le care monde: trois essais de psychologie sociale*. Perspectives du care. ISSN 2556-9511. Lyon: ENS Éditions.

Quijano, Aníbal. 1991. América, el capitalismo y la modernidad nacieron el mismo díaILLA.

Rolnik, Suely. 2018. *Esferas dansurreição. Notas para uma vida não cafetinada*. Sao Paulo: N-1 Ediçoes.

Salamanca González, María Grace. 2020. Éticas del cuidado, decolonialidad e interculturalidad. *Revista RedBioética*, June.

———. 2021. *Soin et résistance en anthropocène: Une éthique-esthétique du care décoloniale*. Lyon: Université Lyon- Instituto Nacional de Antropología e Historia.

———. 2023. *Esthétiques du care pour l'anthropocène*. À Partir de l'Anthropocène. École urbaine de Lyon, Cité Anthropocène, Éditions Deux-Cent-Cinq.

Salamanca González, María Grace, and Paul Hersch Martínez. 2019. La ética del cuidado desde la epidemiología sociocultural: hacia una ética contra-hegemónica presented at the Congreso Nacional ALAMES México, October 24, Zacatecas, México.

Santos, Boaventura de Sousa. 2009. *Una epistemología del Sur : la reinvención del conocimiento y la emancipación social*. Buenos Aires: Siglo XXI, Consejo Latinoamericano de Ciencias Sociales.

———. 2018. *Na Oficina do Sociólogo Artesão aulas 2011–2016*. São Paulo: Cortez Editora.

Steffen, Will, Katherine Richardson, Johan Rockström, Sarah E. Cornell, Ingo Fetzer, Elena M. Bennett, Reinette Biggs, et al. 2015. Planetary boundaries: Guiding human development on a changing planet. *Science* 347: 1259855. https://doi.org/10.1126/science.1259855.

Tronto, Joan C. 1993. *Moral boundaries : A political argument for an ethic of care*. New York: Routledge.

Chapter 4
Between Care and Justice: A Reinterpretation of the Debate in Light of (Inter)national Constitutionalism

Lucia Re

Abstract The framing of the ethics of justice as opposed to the ethics of care has nurtured a three-fold misunderstanding. On one hand, a certain interpretation of Carol Gilligan's *In A Different Voice* emphasized this opposition, even if Gilligan's later writings and other scholars' work have opened up the possibility of delineating a dynamic and potentially harmonious relationship between the "moral voices" of justice and care. On the other hand, the ethics of justice has been identified primarily with John Rawls' neocontractualism. Finally, this ethics of justice and the language of rights have often been confused. The second school of the ethics of care, i.e. the "political theory of care", has reevaluated the importance of the ethics of justice, law, and rights. These scholars have enriched the debate on the relationship between care and justice. The views of some European scholars, particularly Italian ones, further refined this perspective. However, I argue that it is possible to take a further step, starting from interpreting law as a social practice whose reasons for existence and content depend on the attitudes of the interpreters, attitudes which are forged in historical contingencies. In today's constitutional democracies and within international and regional bodies committed to protecting human rights, this practice is oriented by the belief system developed in (inter)national constitutionalism since the second half of the twentieth century. This is a political and legal project based on the principle of equality and recognition of the dignity of every human being. (Inter)national constitutionalism has enriched the liberal legal tradition by theorizing the subject of law as embodied and by recognizing its ontological vulnerability and situated needs. Care, too, is a social practice, and I argue that it is oriented by similar beliefs. The "political theory of care" is thus a valuable ally for (inter)national constitutionalism, as it contributes to innovating and defending it by making visible subjects and needs that have been neglected until now and developing a set of social practices and molecular forms of resistance. In turn, (inter) national constitutionalism represents an inalienable legacy to draw on for achieving a "caring democracy", since it has prepared the institutions and guarantees that care

L. Re (✉)
Università degli studi di Firenze, Florence, Italy
e-mail: lucia.re@unifi.it

J. Braga (ed.), *Ethics of Care*, Advancing Global Bioethics 20,
https://doi.org/10.1007/978-3-031-73608-7_4

can now strengthen and re-signify. Together, they can fight off the neoliberal and conservative attacks that aim to subvert the "political project of equality".

Keywords Ethics of care · Political theory of care · Care · Caring democracy · Constitutionalism · Ethics of justice

1 Introduction

Only recently has the philosophy of law and, more generally, legal reflection begun to approach the ethics and political theory of "care". In this context, the framing of the "ethics of justice" or "rights-based ethics" as opposed to the "ethics of care" has long nurtured a three-fold misunderstanding. On one hand, a certain interpretation of Carol Gilligan's and Nel Noddings's work emphasized this opposition. On the other hand, the "ethics of justice" has been identified primarily with Kantian theories and, among contemporary philosophies, with the neocontractualism of John Rawls. It was thus implicitly assumed that these approaches encompassed all meanings of the concept of justice. And finally, the "ethics of justice", law and rights have often been confused.

The "second school" of ethics of care that has been developing since the 1990s, and some legal feminist scholars engaging with it such as Jennifer Nedelsky, have reevaluated the importance of the "ethics of justice", law and rights. However, the vision of law and rights presented in the work of these Anglo-American scholars appears simplified. The views of some European scholars, particularly Italian ones, help to add complexity to the debate on the relationship between "care" and "justice". Nevertheless, what these analyses do not capture is the deep affinity that exists between the perspective of care and the core of (inter)national constitutionalism[1] that has been developing since the second half of the twentieth century and affects current conceptions of law and rights.

I argue that law should not be interpreted as a given system of positive norms, nor as a set of values that comply with a metaphysical order. Rather, it is to be understood as a social practice whose reasons for existence and content depend on the attitudes of the interpreters, attitudes that are forged in historical contingencies. Today, this practice is oriented, in constitutional democracies and within international and regional bodies committed to the protection of human rights, by the values and, it could be said, the "symbolic order" that have been developing within (inter)national constitutionalism. This is a political and legal project based on the principle of equality and the recognition of the dignity of every human being that has enriched the liberal legal tradition by theorizing the subject of law as an embodied subject and by recognizing its ontological vulnerability and situated needs.

[1]This phrase was coined by Tecla Mazzarese (2018) to underline the national and international character of post-WWII constitutionalism.

This chapter retraces the debate on "care" and "justice". Moreover, suggesting an interpretation of law as social practice, it shows the many affinities between (inter) national constitutionalism, especially in light of its contemporary developments, and the "political theory of care". Finally, it argues that these developments—and the very culture of (inter)national constitutionalism—are now endangered both by a conservative backlash and by the less explicit attack of neoliberal ideology. The "political theory of care" is thus a valuable ally for (inter)national constitutionalism, as it contributes to innovating and defending it by making visible subjects and needs that have been neglected until now and developing a set of social practices and molecular forms of resistance. In turn, (inter)national constitutionalism represents an inalienable legacy to draw on for achieving a "caring democracy" (Tronto 2013), since it has prepared the institutions and guarantees that care can now strengthen and re-signify. Together, they can fight off the neoliberal and conservative attacks that aim to subvert the "political project of equality" (Re 2022a) that constitutes the very basis of (inter)national constitutionalism.

2 Care Versus Justice and Care Versus Law

The development of the "ethics of care" since the 1980s has given rise to a wide-ranging debate on the divergence between this ethical perspective and what is known as the "ethics of justice". Although the distinction between the two is often presented in general terms, the criticism developed by the scholars proposing an ethics of care has primarily targeted Kantian ethics and, through these, liberalism. In particular, advocates of the ethics of care have also directed their critiques at some contemporary normative liberal and neo-contractualist conceptualizations of the ethics of justice, particularly John Rawls' theory of justice (Rawls 1971).[2]

From here, however, the critical analysis has been extended—sometimes explicitly, more often implicitly—to the language of rights, in some cases involving law in its generality. There has thus been a double semantic and conceptual shift from the "ethics of justice" to "rights-based ethics" and from this to law.

Summarizing the elements common to the main ethics of care theories, one can say that these mostly present care ethics as an ethics founded on awareness of the interdependence and vulnerability of human beings and therefore attentive to relationships, concrete situations, "singularities", and the consequences of actions. They usually classify the "ethics of justice" or "rights-based ethics", instead, as an ethics of convictions, universalist and centered on an atomistic conception of the subject. The subject of "rights-based ethics" is considered to be neutral, bodiless and oriented towards fulfilling its own self-interest. It relates to others according to the synallagmatic logic of the contract.

[2] For a survey of this debate see *ex multis* Brotto (2013); Casalini and Cini (2012); Paperman and Laugier (eds. 2011); and Held (ed. 2018).

This conceptualization of the subject of "rights-based ethics" has favored a shift of the critique from such ethics specifically to critiquing law *tout court*, as it views the subject of ethics as coinciding with the legal subject conceived by the Enlightenment that became the cornerstone of the legal order designed by nineteenth-century European codifications. To posit an opposition between "ethics of care" and "rights-based ethics", therefore, also means to criticize the way the legal order has been shaped in liberal states.

As mentioned above, critique also focuses on the contractualist paradigm common to both modern liberalism and contemporary neo-contractualist theories, a fundamental paradigm not only for political theory but also for legal theory. According to this reading, from Hobbes to Rawls the fiction of the social contract has obscured the deep ties between persons, asymmetrical relations of dependence, the community dimension, and differences among persons.

On the level of legal theory, the criticisms posed by ethics of care advocates are, therefore, part of the broad debate inaugurated first by Marxist theories and then by critical legal studies—including feminist legal studies—that have denounced the supposedly neutral modern legal subject on the grounds that it is actually a specifically male, western, propertied citizen.[3]

Legal feminism—in its various declinations—has brought to light the "sexual contract" (Pateman 1988) underlying the social contract, showing the hierarchies that are hidden behind the screen of formal equality. This debate has allowed jurists to "widen the gaze" (Loretoni 2014) so as to recognize long-neglected subjects while proposing a deconstruction of the theory and language of rights.

Feminism—and legal feminism—encompasses many diverse voices. The ethics of care—especially the so-called "first school"—has been considered part of the "second wave" feminism that surfaced in the 1970s. It has been considered in some respects to overlap with Italian and French feminism "of difference" whose positions have been characterized by intense skepticism about the allegedly emancipatory effect of rights and a distrust of the legal system. This attitude of suspicion toward the legal system, considered the medium through which the rule "of the strongest" is institutionalized and legitimized, is certainly not exclusive to feminist approaches, however; indeed, it is a line of thought that has run through Western philosophy since antiquity and found significant expression in Marxist theory.

Feminist legal theory, particularly during "second wave", focused on the sexist character of law. Whereas "first wave" thinkers adopted the emancipationist approach of claiming the same rights for women as men, "second wave" legal theory sought new paths for asserting women's subjectivity and freedom through women's politics, personal relationships and consciousness-raising. The focus here is not an exploration of the differences between the various currents of thought comprising "second wave" feminism, although these differences are much more significant

[3] For a discussion of critical legal theories cf. e.g. Bernardini and Giolo (eds. 2017) and Bauman (2021).

than some critics would seem to believe.[4] What is important here is that these stances—including the ethics of care—have emphasized the need to change the ethical, political and legal paradigm forged in modernity and break out of the restrictive enclosure of the individualistic conception of the subject. In this perspective, the contemporary "age of rights" (Bobbio 1990) has often been interpreted as the era in which the social bond is being eroded by the prevalence of individualism and social action aimed at maximizing personal interests, an attitude which generates conflicts. Observers have also underlined that the legal system entrusts the settlement of these conflicts to the state, in keeping with the typical modern image of justice wielding a sword,[5] instead of fostering multiple sites of dialogue and social mediation, which would be preferable.

3 Reconciling Care and Justice

The early stages of the ethics of care debate posited a rigid separation between ethics of care and ethics of justice due especially to the reception of Carol Gilligan's book *In A Different Voice* (1982) and Nel Noddings's book *Caring: A feminine approach to ethics and moral education* (1984), two works that presented care and justice in stark contrast. The "second" ethics of care school has challenged this rigid separation, with some influential authors beginning to reevaluate the importance of justice and rights, proposing some possible reconciliations between the ethics of care and rights-based ethics. A more open attitude toward the role of justice, law and rights was notably expressed by two leading exponents of the "second school" of care ethics, Virginia Held (2006; ed. 2018) and Joan Tronto (1993, 2013).

Held (2006: ch. 4) argues that it is misguided to oppose care and justice. Moreover, the ethics of care as a feminist ethics cannot underestimate the importance of equality and women's rights. She points out that leading Anglo-American feminist legal theorists from Catharine Mackinnon to Carol Smart, Patricia Williams and Frances Olsen have indeed demonstrated the limits of patriarchal law and warned women against relying on state power to protect their freedoms and bring about change; at the same time, however, they have also emphasized "how rights cannot be replaced by what an ethic of care alone would provide" (Held 2006: 67). Rights are thus tools to be used. They must, however, "be viewed in the context of social practices rather than in the abstract" (Ibid.) so that "they can effectively express the aspirations of a social movement [...]" (Ibid.). The point, then, is not to reject the ethics of justice and language of rights but to downgrade their importance, placing them within a pluralist conception of ethics as one of the multiple possible ways to sustain moral choices and political decisions.

---

[4] I refer in particular to Fraser (2013). For a discussion of these different positions feminism has developed regarding law, see Re (2019).

[5] For a critical analysis of the historical evolution of the image of justice see Simone (2015).

According to Held, a rights-based ethics may prove *unsuitable* in some contexts, including contexts that are still considered to be handled most effectively through justice and rights. Even within the legal field where justice and rights should generally have priority, various issues can reveal their limits: this often occurs, for instance, with family law and measures to combat gender-based violence.[6] In many cases, taking into account the context and logic of the relationships involved shows that "other moral considerations should play a larger role" (Held 2006: 68).

Hence, justice and care are both needed. They should be regarded not as alternative paradigms but rather as mutually complementary perspectives. Held, however, grants priority to care, considering it "the most basic moral value" (2006: 71) since life would be impossible without it. In her view, care is also a "more inclusive value" (2006: 72) because it is closer to human experience. The ethics of care is based on the observation of the vulnerable and relational reality of human beings. By outlining a society composed of sovereign individuals who are independent of each other and imagining it is possible to identify a neutral universal standpoint, justice instead builds a fiction for political and social ends. The "picture" of justice should therefore be embedded "in the wider tapestry of human care" (Ibid.).

In Tronto's (1993) view, the ethics of care is likewise not opposed to the ethics of justice. She aims to free care from the artificial boundaries set by the separation of ethics and politics, the moral point of view understood as distinct from concrete moral problems in daily life, and private "vs" public spheres. Taking as her horizon of reference a liberal and pluralistic democracy, she holds that care should be placed at the center of politics. By valorizing care, Tronto (2013) aims to transform this horizon into a "caring democracy". Like Held, she believes that justice without an idea of care is incomplete (Tronto 1993: ch. 6). At the same time, care must be embedded within a theory of justice lest it become paternalistic/maternalistic and remain limited to only some kinds of relationships.

Equality is at the center of Tronto's concerns. Although her focus is on people's needs and both personal and institutional responsibilities, she makes it clear that the rights of care-receivers must absolutely be taken into account. The point is not to reject rights, therefore, but to make them effective and to foster democratic discussion about them. Indeed, such discussion must never go so far as to call into question certain rights that Tronto (2013: 153–154) clearly considers fundamental: 1. the "right to receive care", 2. the "right to care", and 3. the right "to participate in the public process by which judgments are made about how society should ensure the first two premises".

The political theorist clearly states that care is not to be understood as a good to be distributed, but rather as an "activity in which citizens are constantly engaged" (Tronto 2013: 154). There needs to be public debate about both the best way to support this activity and the quality of care, with care givers and care receivers all included in the dialogue. It must also be possible to establish "a right not to care" for people in specific situations.

---

[6] Cf. e.g. Re (2021a).

Tronto's perspective is important because it shifts the focus from ethics to politics,[7] showing the possibilities for immediate action in contemporary democracies. It not only combines care and justice and care and rights, but also highlights that they are all contextual and should therefore be subject to democratic discussion by all stakeholders. In this way, care also becomes a tool for improving democracy by making it more inclusive and changing the way public institutions, communities and individuals act.

Finally, other work with special significance for the purposes of my discussion is that of Jennifer Nedelsky (2011), a legal theorist who, while not identifying with the ethics of care, is very adjacent to this perspective.[8] Nedelsky goes beyond the considerations traced so far to theorize a true reshaping of the legal field from a relational ethics standpoint. Her attention is directed in much greater detail not only at law but also at constitutionalism. Although not declared as such, her approach is that of American legal realism, looking at law in action and paying particular attention to judges' decisions.

Nedelsky deliberately chooses to limit her analysis to North America, referring to the different legal systems of Canada and the United States. From this point of view, she formulates a strong critique of both the power of judges—viewed as technocrats acting outside of real democratic control—and, for the same reasons, judicial review. This critique is highly problematic, since judicial review is the cornerstone of constitutionalism. Indeed, it protects fundamental rights from the "tyranny of the majority" (Tocqueville 1835–1840: vol. 1, part 2, ch. 7), obliging the law to comply with rights.

In my view, this critique derives from a simplified understanding of law. Although Nedelsky's focus on law is deeper than that of the aforementioned care ethics scholars, it likewise ends up neglecting the complexity of the phenomenon that is law. The impression is that adopting a strictly realist approach to law limits her reading of constitutionalism and diminishes the scope of her interesting theoretical proposal of reinterpreting the entire legal field by shifting the focus from the subjects, goods, abstract legal institutes etc. shaped by modern liberal legal theory to "law's relations". This reconceptualization brings significant changes in the way law and rights are conceived, in the democratic debate about them, and in the work of judges and legal practitioners. However, I think that it would benefit greatly from engaging with European constitutionalist and legal-philosophical reflection, particularly work inspired by legal hermeneutics. Indeed, it is this perspective that, as I will argue, allows us to enrich not only the democratic debate on rights and law but also legal interpretation, making the judicial method the hinge between the instituting and the instituted dimensions of the legal order.[9]

[7] This is why she prefers to call her approach by the name "political theory of care" instead of "ethics of care".

[8] I would like to thank Brunella Casalini for reminding me of the importance of Nedelsky's work for the arguments I am developing here.

[9] I have posed some consideration on this topic in Re (2022b), also commenting on Ciaramelli (2021).

4 A View from Europe

All of the perspectives more or less directly referable to the second ethics of care school that I have outlined so far are of great importance in that they suggest reconciling the ethics of care and rights-based ethics and stress the relevance of rights, especially for the purpose of countering asymmetrical power relations and avoiding forms of paternalism/maternalism. In most cases, however, they assume a continuity between modern and contemporary times and so underestimate the distinct caesura produced by the passage from the modern "legislative state" to the contemporary "constitutional state".[10] I believe, instead, that it is useful to highlight this shift while at the same time better situating the analysis in the European context, where the history of (inter)national constitutionalism has been linked to the development of the welfare state and the principle of substantive equality has been positioned at the foundations of the new conception of the rule of law. Indeed, many of these theories seem to be biased by a North American point of view that overlooks the outcomes of European legal—and legal-philosophical—reflection.

In the Italian philosophical debate, the need to reconcile the ethics of care and rights-based ethics has been emphasized by some authors with an ethics of care orientation. They have both retraced the North American debate and proposed new keys to interpretation. Vittoria Franco, for example, has argued that it is necessary to formulate a new "grammar of justice" (Wolgast 1987) "capable of interpreting the complexity of relationships without resorting to preconstituted grids and without, however, refusing legitimacy to claims concerning rights" (Franco 1996: 198), since "rights ethics is a prerequisite for women's ethics as well" (Ibid.).

Anna Loretoni (2018: 88) suggests that "responsible care is a dimension necessary to the very effectiveness of rights, since it represents a move toward others and their needs, a paying attention and listening to others through an empathic practice that thus takes on relational value and moral relevance". From this perspective, "configuring care as a value of moral and political life capable of broadening our horizons does not mean abandoning the language of rights, the significant limits of which must also be acknowledged" (Loretoni 2018: 89). Loretoni (Ibid.) suggests that "feminist morality could also make use of general principles, avoiding the risk of being reduced to a relativist ethics of 'case by case'. These principles, however, will not be grounded in purely rational subjects, but will be structured in such a way as to be adapted to concrete cases; we should in fact arrive at formulating them through experience, made up of empathic and caring dimensions, of trust-based relationships". For the author, it is "thus not a matter of applying these principles mechanically, along a line from the general norm to the specific case, but of 'putting them into practice', modifying and adapting them to the specificity of contexts, including through a 'return' movement that engages with the concreteness of situations to try to redefine the general starting-point features" (Ibid.).

[10] For a distinction between the two, see Ferrajoli (2002).

Finally, the most articulate proposal for combining care and justice in the Italian philosophical debate was made by Elena Pulcini in her last book, titled precisely *Tra cura e giustizia: Le passioni come risorsa sociale* (2020). Here, she argues that care and justice are "two different 'regimes of action' with equal dignity and efficacy" (Pulcini 2020: 35). The former "takes on the problem of defending equality (against oppression and inequality)", while the latter "takes on the problem of protecting relationships" (Ibid.).

The concept of justice that Pulcini looks to is not the one delineated by Rawls' theory, however, but rather the one proposed by Amartya Sen (2009) in his call for "starting from concrete realities and situations of injustice rather than from the definition of an ideal model of justice" and invitation to "focus the question of justice first and foremost on the verification of concrete social achievements, and, second, on the comparative consideration of the steps through which justice can be promoted (rather than on the definition of perfectly just agreements)" (Pulcini 2020: 56). This is therefore a justice that admits "the possibility of partial and imperfect solutions to be explored from time to time, renouncing claims to completeness and the logic of 'all or nothing' [...]" (Pulcini 2020: 57). According to the author, "The merit of this theoretical perspective lies in the fact that, unlike ideal theories of justice, it adopts a more realistic and pragmatic approach, concerned primarily with recognizing and addressing the injustices that exist here and now, rather than identifying them based on an abstract model of a perfectly just society" (Ibid.). Justice viewed in this light is thus, we might say, a search for equality to be measured against the concrete reality of social relations. Care and justice share a common motivation and, more importantly, are two ways of responding to vulnerability. Both have to do with the need to relate to others, not only those near us but also those far away.

As I will discuss in more depth here below, this movement from the concrete to the abstract and vice versa suggested by both Loretoni and Pulcini is typical of the legal method and judicial interpretation in particular. In a sense, it could be said that "legal justice" in constitutional states has precisely this character.

The theoretical positions of the ethics of care "second school" show how much more articulated the debate has become over the years. In this framework, the considerations developed by Italian authors are less affected by the North American philosophical and legal approach. Even in their work, however, the notions of law and rights appear simplified. To support these theoretically and politically relevant positions, therefore, I believe it is important to highlight the complexity of the legal dimension. To this end, it must be recognized first of all that the normative ethics of justice and the language of rights are not the same thing, and even less overlapping are these and what one can consider law. Rights-based ethics is usually linked to a reductive view of fundamental rights, while the status of fundamental rights is instead the subject of a broad legal-philosophical debate.

Secondly, it is appropriate to distinguish the idea of justice that emerges from political-philosophical reflection on normative ethics of justice from the one that derives from legal reflection on justice.[11] The two have important points of contact, but they differ in terms of some salient features. We have to distinguish between them; only then will we be able to look at them jointly obtaining a richer notion of justice.

In the following sections, therefore, I argue that in the framework of contemporary (inter)national constitutionalism, care and law/rights are not only compatible but deeply linked. The perspective of care in which care is freed from essentialism and also from communitarianism, thereby leaving space for relational autonomy—i.e. the autonomy of relational selves who are not independent and always in-control, but are still capable of creative interaction[12]—allows us to account for the conception of the legal subject, fundamental rights, the role of the state and, more generally, law in the framework of contemporary (inter)national constitutionalism, especially in light of its most recent developments. I am thinking in particular of the form of constitutionalism Ruth Rubio-Marín (2022) has called "transformative", representing as it does an evolution of post-WWII (inter)national constitutionalism in the direction of subverting the original gender order and having been developed by feminist theory and politics.

Finally, explaining and strengthening the link between law, rights and care can contribute not only to broadening our critical gaze, but also to defending the political and legal project of (inter)national constitutionalism from the attacks it is currently facing from both neoliberal and "sovereignist" forces,[13] and to developing it in a direction capable of facing the challenges posed by the present and future.

5 Law as Social Practice

As I have argued above, the vision of law deployed in the debate on "care and justice" is largely a simplification. It appears to be identified either with the natural law theory that associates rights and moral values, thereby basing the validity of law on adherence to a metaphysical order, or with a legal positivist conception according to which law is reduced to a set of formal norms adopted by the state.

Interpreted in this way, the justice-law nexus becomes either a question of an exclusively philosophical order that consists in identifying the material content of justice regardless of the "law in action", or a question of mere respect for legality.

As mentioned, even Nedelsky's proposal emphasizing the relational character of rights offers a simplified view of rights, defining them as "collective decisions about the implementation of core values" (Nedelsky 2011: 233) and focusing on decision-making processes at both the political and judicial levels. This perspective does not

[11] Cf. e.g. Del Bo (2022).

[12] See Nedelsky (2011) and Mackenzie et al. (eds. 2014).

[13] Cf. e.g. Re (2022a) and Re (2021b). I use the term "sovereignist forces" to refer to nationalist right wing political organizations and parties.

succeed in capturing the life of law in its fullness, even if it sets off from the observation that law is conceivable only within an intersubjective dimension.

Nedelsky emphasizes the ability of law to structure relationships. However, she frames law as an instrument of government, neglecting its character as a social practice. Moreover, although she argues for abandoning some ways of thinking about law, criticizing in particular the boundary metaphor she sees as structuring Anglo-American law, she does not seem to fully appreciate the importance of the symbolic order that is attached to law. While devoting attention to problems of a linguistic and epistemic nature, she does not go so far as to use these to interpret the life of law. This is not merely a theoretical issue, for it is precisely this blind spot in her theory that leads her to a radical critique of judicial review, instead entrusting democratic deliberation with the task of protecting—and redefining—the fundamental values of the legal system. This position undermines the stabilizing function of law. On the contrary, the effort of contemporary constitutional systems lies precisely in combining this need with the diverse demands for change and recognition that continually emerge in pluralistic societies and cannot always find a voice in democratic debate. This is even more true for today's representative democracies where, as Nedelsky herself acknowledges, it is difficult for some people and groups to make their voices heard.

The idea that law is a social (and thus relational) practice is, in my opinion, best brought into focus by legal hermeneutics, a lens that manages to fully grasp the core of contemporary (inter)national constitutionalism. Law has a horizontal dimension that is not ancillary but constitutive. It regulates relationships between people and contributes to establishing mutual expectations (cf. Greco 2021). A similar character has been attributed to justice in philosophical-legal thought, where it has been recognized that justice "[...] inserts individuals into a coordination in which legitimate claims are determined for each person by their relationship with others" (Pastore 2012: 262). Therefore, justice carries with it a "metegoistic" dimension that renders it open to sociality (Pastore 2012: 262–263). Even without adopting a specific notion of justice and agreeing with the realist vision proposed by Alf Ross (1958) according to which it can take on highly diverse meanings depending on the culture of which it is an expression, one can hardly deny its relational character.

Moreover, precisely because the notion of justice is variable in its content, it is appropriate to examine how it is formulated and defined in the legal and political culture of contemporary (inter)national constitutionalism. In today's constitutional democracies, the idea of justice refers to the symbolic order forged by constitutionalism and translated into formal norms. It is thus linked to certain legal principles; these principles are not, however, crystallized once and for all in normative texts. They are instead specified, redefined, and enlivened by interpretation in each instance as part of a diachronic process that is continually evolving.

In this framework, human rights can be interpreted as justice criteria related to the expectation of eliminating socially produced suffering (Pastore 2022: 6). Such rights were not handed down from above—they are the fruit of political, social and legal struggles. They are unthinkable except in connection with a society that is able to support them and help them evolve through the political and legal processes of claiming.[14] The relational dimension emphasized by Nedelsky is, therefore, already acknowledged in the culture of contemporary (inter)national constitutionalism. It is then primarily a matter of highlighting what already characterizes the life of law, looking at it not only as a set of directives addressed to citizens by the centers of decision-making power but also, and more importantly, as a social order that is connected to a collective imaginary.

In particular, it should be recognized that (inter)national constitutionalism has innovated with respect to the tradition of modern liberal law precisely by virtue of organizing the legal and political system to revolve around not the Enlightenment-inspired abstract subject but the person, that is, situated subjects recognized in their different phases of life, with all their differences and the varying economic and social conditions in which they find themselves.[15] It is specifically this principle derived from personalism that requires situated decision-making (Pastore 2022: section 3): interpretive practices constantly carry out further contextualization.

The dignity of the person, not only formal but also substantive equality, and solidarity are among the cardinal principles of contemporary constitutionalism. They spring from the recognition of human vulnerability and look at the subject as a person existing in relation to others (Pastore 2018).

This perspective allows us to identify the three essential characteristics of justice today: its being *ad alterum*, its having as its object a *debitum*, and its connection with the principle of equality.[16] Law gives a concrete shape to justice understood in this way. It does not coincide with formal legal norms nor with a set of established values, but rather is a combination of texts, meanings, arguments, and social practices (Pastore 2012: 272–273). It is "a plural collective enterprise, undergoing continuous development" in which "rules, principles, institutes, procedures, institutions, normative attitudes, appreciations of value, interpretive practices, reasoning, conceptual constructions, and systematizations" (Pastore 2012: 273) come into play. The life of law is therefore continuous translation, adaptation, and mediation. And it is in this practice that "singularities" emerge.

Jurisdictional activity can also be read through this lens not as a mere exercise of the state's power to decide who is right and who is wrong, but as a moment of mediation between the interests and expectations of the parties in conflict.[17] This mediation should be understood as aimed at both preserving relationships that are considered good and correcting or dismantling harmful relationships.

The various actors involved contribute to the interpretive practice of law. Interpretation cannot be thought of as an act of solipsistic decision-making on the part of judges;[18] the legal culture of constitutionalism is instead distinguished by

[14] Regarding the political and social support for constitutionalism, see the extensive work of Gaetano Azzariti. I have tried to stress the relationship between conservation and social change with reference to constitutionalism in Re (2022b).

[15] See Ciaramelli (2013).

[16] Pastore (2012: 263). On this topic see also Bobbio (1985).

[17] See the reading Ciaramelli (2017) offers of the conflict between Antigone and Creon.

[18] Cf. Santoro (2008: ch. 4).

responsiveness.[19] This view does not deny that the legal system also gives rise to vertical relations and provides for the use of coercion, but it recognizes that the system's main role lies in coordinating social relations (Greco 2021: 147).

In this respect, law and justice can be considered similar to care, as care is also a social practice. Not only do they share this quality, but there is also a deep relationship between law, justice and care: law and justice must be the object of care[20] in order for the legal system to function as an instrument of social pacification and means for coordinating citizens' expectations. Rights also need care so that they may be recognized, extended, claimed at a political, social and judicial level, read in their interconnections, made effective through a system of jurisdictional guarantees and social protection.

6 Taking Care of Constitutionalism

The interpretation of (inter)national constitutionalism I have laid out here is of course not undisputed. Indeed, I have read it in light of theories of vulnerability and care. What I intend to argue, however, is that the basic features of contemporary constitutionalism not only allow but actually encourage this kind of reading. And such an interpretation is all the more important today, at a time when neoliberalism—together with its unsuspected ally, sovereignism[21]—is "undoing the demos" (Brown 2015) and undermining our constitutional states by turning them into "careless states" (The Care Collective 2020). Indeed, we are at a crossroads between the possibility of preserving the roots of post-WWII constitutionalism and even growing them further in an anti-paternalistic direction, and the risk of a powerful backlash aimed at overwhelming the very architecture of constitutional states.

Rubio-Marín (2022: 20) has identified four forms of constitutionalism:

1. "exclusionary gender constitutionalism, that manifested from the birth of written constitutionalism until post-World War II". Indeed, modern constitutionalism allowed women only half-citizenship, depriving them of fundamental rights that, while presented as universal, were mostly granted only to men. Specifically, the institution of marriage guaranteed husbands legal power and control over women, granting legal legitimacy to a relationship of subordination asserted to be in

---

[19] Cf. Greco (2021: ch. 6) and Fineman (2010).

[20] Here, I employ the notion of care in the broad sense theorized by Berenice Fisher and Joan Tronto (1990: 40): "On the most general level, we suggest that caring be viewed as a *species activity that includes everything that we do to maintain, continue, and repair our 'world' so that we can live in it as well as possible*. That world includes our bodies, ourselves, and our environment, all of which we seek to interweave in a complex, life-sustaining web".

[21] "Sovereignist" parties pose as advocates for citizens against immigrants, against globalization and, in Europe, against the technocracy of the European Union. When they govern, however, they often end up adopting neoliberal economic policies and fiercely opposing redistributive measures. For a discussion of the alliance between neoliberalism and sovereignism cf. Serughetti (2021).

accordance with the nature of the two sexes. This form of constitutionalism enacted the "constitutional embedding of the breadwinner family order" (Ibid.: 132);

2. "inclusive gender constitutionalism", which began to recognize women's rights by equating them with men but only in some respects, at the same time maintaining a strict division of care labor and gender roles. In Europe, this form of constitutionalism allowed for the protection of motherhood and recognition of related rights, but ended up creating a "maternalist" model that disadvantaged women's participation in public life and the labor market;[22]
3. "participatory gender constitutionalism", linked to developments in feminism in both Western countries and internationally and the demand that women be able to take active roles in public life and the production of norms;
4. "transformative gender constitutionalism", which aims at "the disestablishment of patriarchal constitutionalism's foundational gender order" (Rubio-Marín 2022: 21) and places women's reproductive rights and care at the center of the constitutional agenda. It also goes beyond a women's rights perspective to demand the recognition of different gender identities.

This list of different forms only partially corresponds to the historical phases constitutionalism has undergone from the beginning of the modern age to the present. Initially it mainly concerns European and North American countries, but then extends to consider other areas of the world in which constitutionalism has been developing. Indeed, it is precisely those countries with more recent constitutions—particularly in South America and South Africa—that offer the most representative models of transformative constitutionalism today. Moreover, phases and trends partly overlap with each other. Finally, this reconstruction is intended not only to examine women's positions and rights within constitutionalism but also, and most importantly, to reveal how gender issues in particular are constitutive of different forms of constitutionalism.

Although Rubio-Marín cautions that her proposed four forms of constitutionalism do not correspond to precise periods or outline a trajectory from backwardness to emancipation, one cannot help but notice how transformative constitutionalism represents the full development of the principles that (inter)national constitutionalism had posited in the aftermath of WWII. In particular, it could be argued that (inter)national constitutionalism began in form 2 ("inclusive gender constitutionalism") but continues to develop into form 4 ("transformative constitutionalism"), even if such development is not linear or uninterrupted. At the same time, "transformative constitutionalism" reflects a new social order that has now taken shape, connected to a symbolic imaginary that rejects sexism and heteronormativity as ordering principles of social relations. It is composed of a range of social practices including normative production, legal claiming, legal interpretation, etc. It thus reinterprets universalism starting from the recognition of differences and defends a substantive

[22] Regarding the difficulty of overcoming this model even today in the Italian constitutional system, cf. Magneschi (2022).

conception of equality, drawing attention to the effectiveness of rights. Indeed, it is only by positioning the issue of care and social reproduction at the center of the constitutional agenda that the core idea of (inter)national constitutionalism, the protection of human dignity and equality, can be implemented.

Paradoxically, just when contemporary constitutionalism is nearest to achieving this goal it is being challenged by a fierce backlash that undermines its very personalist core and universalist inspiration. The principles affirmed in it are openly questioned at times but, even more frequently, they are drained of vitality through the spread of an extreme individualistic careless culture and the erosion of the welfare state.

As Wendy Brown (2015) has pointed out, this attack is also made by transforming the conceptualization of law and the way it operates. Orsetta Giolo (2020: 33) has observed that neoliberal law abandons the idea of the person typical of (inter) national constitutionalism and attacks the project of equality, contributing to the consolidation of social hierarchies. Indeed, it operates according to a binary logic: economically and socially stronger subjects are able to make use of the multiple options offered by global law and various forms of soft law that facilitate their operation, while "non-paradigmatic subjects" who do not conform to the dominant order live in a different "legal environment", an "authoritarian" one where personal freedoms are subject to serious limitations and fundamental rights are mostly ineffective.

A conflict is playing out between what we might call the "equality project" characteristic of (inter)national constitutionalism and the "inequality project" promoted by neoliberal capitalism (cf. Re 2022a). This clash also involves the law: the reaffirmation and even full development of constitutionalism's values come up against a neoliberal twist in the law that distorts the very assumptions of contemporary constitutionalism. In this scenario, the position of gender rights—and gender issues more generally—is extremely uneasy. Gender rights and identities are denied by reactionary movements focused on rejecting gender equality and women's reproductive autonomy as compensation for having lost sovereignty in other spheres of social life, from geopolitics to the economy, labor, health etc.[23] And at the same time, gender rights and identities are subject to neoliberal appropriation that turns them into marketable goods. In particular: so-called "civil rights" are increasingly opposed to "social rights"; formal equality replaces the quest for substantive equality and plural participation in democratic life; and the value of liberty is reinterpreted as mere freedom of choice among market offerings (cf. Facchi and Giolo 2020).

The perspective of care is thus key for escaping this two-part 'commandeering' of rights precisely because it allows rights to be removed from the logic of the market, reinterpreting them as claims embedded in social practices in continuity with the values promoted by (inter)national constitutionalism. Care is also a response to the social anxiety generated by globalization. It offers both an individual and

[23] A similar interpretation has been provided by Rubio-Marín (2022, in particular: 701–702).

community response and aims to rebuild the social bonds eroded by neoliberal capitalism (cf. Lynch 2022). Finally, emphasizing the 'everyday practice' nature of care allows us to overcome ideological polarization by bringing the exchange of views and action to concrete issues that can be addressed through mutual listening and recognition.

The political theory of care is a valuable ally of constitutionalism, therefore, not only because—by reflecting on care labor—it can foster transformative constitutionalism, but also because it suggests 'caring together' for democracy and pluralism, renewing institutions and opening up social participation in part by enhancing the emotional dimension of politics. On one hand, it pushes for the full realization of substantive equality by opening up the conflict surrounding the recognition of and responsibility for care work as well as its fair distribution. On the other hand, it emphasizes the political importance of passions (cf. Pulcini 2009, 2021). These passions, liberated and re-signified by a care that becomes "promiscuous" (The Care Collective 2020)—that is, no longer confined within traditionally structured social ties (starting with the monogamy family)—can thus emerge from what Tocqueville (1835–1840, vol. 2, ch. 4) called the small circle of self-interest and take their place at the center of the agora, contributing to defending the "political project of equality" (Re 2022a).

Adopting the perspective of care, as Tronto (2013) argues, means fulfilling and valuing a set of care practices to be cultivated daily, at all levels. These practices involve the legal field and help in shaping the symbolic order and developing a new democratic 'paideia' that erases the separation between the private and public spheres while teaching us how to 'care for passions', cultivating the ethical quality of emotions (Pulcini 2020). This new "paideia" can also help strengthen the "passion" for the Constitution that many contemporary defenders of constitutionalism, from Jürgen Habermas to Luigi Ferrajoli, consider an essential condition for the resilience of constitutional democracies. Indeed, it not only solicits citizens' participation by making them aware of their constitutional rights and duties, but also provides access to a different and deeper understanding (including emotional) of constitutional values.

The road ahead for building caring communities and "caring states" inclined to international cooperation is long and difficult. During this journey, however, we can rely on the legacy inherited from twentieth century (inter)national constitutionalism by committing ourselves to implementing and further developing its promises in the direction of a "caring constitutionalism".

References

Bauman, Richard W. 2021. *Critical legal studies. A guide to the literature*. New York/London: Routledge. Ebook.

Bernardini, Maria Giulia, and Orsetta Giolo, eds. 2017. *Le teorie critiche del diritto*. Pisa: Pacini.

Bobbio, Norberto. 1985. Sulla nozione di giustizia. *Teoria politica* 1: 7–19.

———.1990. *L'età dei diritti*. Turin: Einaudi. English Edition: Bobbio, Norberto. 1995. *The age of rights*. Trans. A. Cameron. Cambridge/Malden: Polity Press.

Brotto, Sara. 2013. *Etica della cura. Una introduzione*. Naples: Orthotes.

Brown, Wendy. 2015. *Undoing the demos. Neoliberalism stealth revolution*. New York: Zone books.

Casalini, Brunella, and Lorenzo Cini. 2012. *Giustizia, uguaglianza e differenza. Una guida alla lettura della filosofia politica contemporanea*. Florence: Firenze University Press.

Ciaramelli, Fabio. 2013. *Consenso sociale e legittimazione giuridica. Lezioni di filosofia del diritto*. Turin: Giappichelli.

———. 2017. *Il dilemma di Antigone*. Turin: Giappichelli.

———. 2021. *L'ordine simbolico della legge e il problema del metodo*. Turin: Giappichelli.

Del Bo, Corrado. 2022. *La giustizia. Un'introduzione filosofica*. Roma: Carocci.

Facchi, Alessandra, and Orsetta Giolo. 2020. *Libera scelta e libera condizione. Un punto di vista femminista su libertà e diritto*. Bologna: Il Mulino.

Ferrajoli, Luigi. 2002. Lo Stato di diritto fra passato e futuro. In *Lo Stato di diritto. Storia, teoria, critica*, ed. Pietro Costa and Danilo Zolo, 349–386. Milano: Feltrinelli.

Fineman, Martha A. 2010. The vulnerable subject and the responsive state. *Emory Law Journal* 60: 251–275.

Fisher, Berenice, and Joan C. Tronto. 1990. Toward a feminist theory of caring. In *Circles of care*, ed. Emily K. Abel and Margaret K. Nelson, 36–54. Albany: SUNY Press.

Franco, Vittoria. 1996. *Etiche possibili. Il paradosso della morale dopo la morte di Dio*. Rome: Donzelli.

Fraser, Nancy. 2013. *Fortunes of feminism. From state-managed capitalism to neoliberal crisis*. London: Verso.

Gilligan, Carol. 1982. *In a different voice. Psychological theory and women's development*. Cambridge, MA: Harvard University Press.

Giolo, Orsetta. 2020. *Il diritto neoliberale*. Naples: Jovene.

Greco, Tommaso. 2021. *La legge della fiducia. Alle radici del diritto*. Rome-Bari: Laterza.

Held, Virginia. 2006. *The ethics of care. Personal, political, global*. Oxford/New York: Oxford University Press. Ebook.

———, ed. 2018. *Justice and care. Essential readings in feminist ethics*. London/New York: Routledge. Ebook.

Loretoni, Anna. 2014. *Ampliare lo sguardo. Genere e teoria politica*. Rome: Donzelli.

———. 2018. Etica della cura e vulnerabilità degli individui. *La società degli individui* 3: 77–90. https://doi.org/10.3280/LAS2018-063006.

Lynch, Kathleen. 2022. *Care and capitalism. Why affective equality matters for social justice*. Cambridge: Polity Press.

Mackenzie, Catriona, Wendy Rogers, and Susan Dodds, eds. 2014. *Vulnerability. New essays in ethics and feminist philosophy*. Oxford: Oxford University Press.

Magneschi, Chiara. 2022. Per una (ri)costituzionalizzazione della cura: tra analisi critica e prospettive normative. *Teoria e storia del diritto privato Special Issue Ombre del diritto*: 1–31.

Mazzarese, Tecla. 2018. I migranti e il diritto ad essere diversi nelle società multiculturali delle democrazie costituzionali. In *Strumenti, percorsi e strategie dell'integrazione nelle società multiculturali*, ed. Ginevra Cerrina Feroni and Veronica Federico, 63–85. Naples: ESI.

Nedelsky, Jennifer. 2011. *Law's relations. A relational theory of self, autonomy, and law*. Oxford: Oxford University Press.

Noddings, Nel. 1984. *Caring. A feminine approach to ethics and moral education*. Berkeley: University of California Press.

Paperman, Patricia, and Sandra Laugier, eds. 2011. *Le souci des autres. Éthique et politique du care*. Paris: Éditions de l'Ecole des hautes études en sciences sociales. Ebook.

Pastore, Baldassare. 2012. Giustizia. In *Atlante di filosofia del diritto*, ed. Ulderico Pomarici, vol. 1, 260–279. Turin: Giappichelli.

———. 2018. Soggettività giuridica e vulnerabilità. In *Vulnerabilità. Analisi multidisciplinare di un concetto*, ed. Orsetta Giolo and Baldassare Pastore, 127–145. Rome: Carocci.

———. 2022. Vulnerabilità, diritto, ragionamento giuridico. *Teoria e storia del diritto privato*: 1–18. special issue.
Pateman, Carol. 1988. *The sexual contract*. Cambridge: Polity Press.
Pulcini, Elena. 2009. *La cura del mondo. Paura e responsabilità nell'età globale*. Turin: Bollati Boringhieri. English Edition: Pulcini, Elena. 2013. *Care of the world: Fear, responsibility and justice in the global age*. Trans. K. Whittle. Dordrecht/New York: Springer.
———. 2020. *Tra cura e giustizia. Le passioni come risorsa sociale*. Turin: Bollati Boringhieri.
———. 2021. Soggetti emozionali. Per aver cura del futuro. *Iride* 34: 267–283. https://doi.org/10.1414/102455.
Rawls, John. 1971. *A theory of justice*. Cambridge, MA: The Belknap Press of Harvard University Press.
Re, Lucia. 2019. Eguaglianza, differenza, diritto. Uno sguardo al dibattito femminista contemporaneo. *About Gender* 15: 1–42. https://doi.org/10.15167/2279-5057/AG2019.8.15.1059.
———. 2021a. Il nodo della violenza: diritto, gender mainstreaming e libertà femminile nel dibattito italiano. *Notizie di Politeia* 143: 59–71.
———. 2021b. Restoration or remodeling? The constitutional state in the postmodern era. *Soft Power* 8: 235–242. https://doi.org/10.14718/SOFTPOWER.2021.8.1.14.
———. 2022a. Difendere l'uguaglianza. Verso la democrazia costituzionale della cura. *La società degli individui* 73: 71–82. https://doi.org/10.3280/LAS2022-07300.
———. 2022b. Affrontare la crisi: il costituzionalismo democratico fra conservazione e creazione. *Etica & Politica* 24: 321–339.
Ross, Alf. 1958. *On law and justice*. London: Steven & Sons.
Rubio-Marin, Ruth. 2022. *Global gender constitutionalism and women's citizenship. A struggle for transformative inclusion*. Cambridge/New York: Cambridge University Press. Ebook.
Santoro, Emilio. 2008. *Diritto e diritti. Lo Stato di diritto nell'era della globalizzazione*. Turin: Giappichelli.
Sen, Amartya. 2009. *The idea of justice*. London: Penguins Books.
Serughetti, Giorgia. 2021. *Il vento conservatore. La destra populista all'attacco della democrazia*. Rome-Bari: Laterza.
Simone, Anna. 2015. *Rappresentare il diritto e la giustizia nella modernità. Universi simbolici, iconografia, cambiamento sociale*. Milan: Mimesis.
The Care Collective. 2020. *The care manifesto. The politics of interdependence*. London/New York: Verso. Ebook.
Tocqueville, Alexis de. 1835–1840. *De la démocratie en Amérique*. Paris: Librairie de C. Gosselin. English Edition: Tocqueville, Alexis de. 1964. *Democracy in America*. Trans: H. Reeve. New York: Washington Square Press.
Tronto, Joan C. 1993. *Moral boundaries. A political argument for an ethic of care*. New York/London: Chapman/Routledge.
———. 2013. *Caring democracy. Markets, equality and justice*. New York: New York University Press.
Wolgast, Elizabeth H. 1987. *The grammar of justice*. Ithaca: Cornell University Press.

Chapter 5
Care for Oneself and Care for Others in Moral Virtues: Surpassing the Dilemma

José Manuel Beato

Abstract The purpose of this chapter is to analyze some aspects of the tension between 'care for oneself' and 'care for others' from an aretological approach to morality. A traditional perspective—based on a metaphysical foundation—may conceive virtue ethics as centered on the affirmation and realization of the self. Conversely, an ethics of care will tend to have an ethical perspective centered on the other. However, a virtue theory centered on a "relational self" that reconfigures its spectrum by considering care as a virtue among other virtues (justice, temperance, courage, etc.) will allow us to overcome the antinomy of self and other centredness and render the opposition between these ethical proposals obsolete. The traditional principle of unity and mutual implication of virtues is reinforced if we see virtues as multiple expressions of care and care as a virtue that implies the whole aretological spectrum. From these assumptions, a new perspective of "moral perfectionism" emerges.

Keywords Care · Virtue · Perfectionism · Good

1 Introduction: Caring Among Virtues

If we define 'care' as encompassing both the disposition to 'care for' and 'care about' others, as well as the ability to 'take care of' them, resulting in caregiving activities, then we are dealing with a complex set of affective and volitional aptitudes. These are coordinated by a kind of practical wisdom that enables actions of ethical significance. As such, this capacity can be considered a form of moral excellence, similar to what is traditionally recognized as a virtue.

As characterized, care is a virtue—a moral disposition developed through practice and habituation—that can rightfully take its place alongside traditional virtues and be integrated into an ethical theory with fundamental aretaic features. The

J. M. Beato (✉)
Institute for Philosophical Studies (IEF), Universidade de Coimbra, Coimbra, Portugal

J. Braga (ed.), *Ethics of Care*, Advancing Global Bioethics 20,
https://doi.org/10.1007/978-3-031-73608-7_5

virtue of care is primarily relational and other-regarding, challenging the moral agent to adopt a more radical, deep, and broad openness to otherness. In virtuous practices of care, the interests of the 'one-caring' and the 'one-cared for' become intertwined, thus challenging the theoretical and practical constitution of a 'relational self.' In this sense, care is a 'relational virtue,' where the excellence of the relationship takes precedence over the excellence of the individual agents involved (Beato 2021).

Emphasizing theoretical contiguity and promoting practical convergence between the ethics of care and the ethics of virtues carries significant theoretical and practical implications. Not only can we conceive of and enact care as a virtue, but we can also view traditional virtues as diverse expressions of care. This assessment, along with the principle of reversibility that underpins it, invites exploration across various domains of applied ethics, such as bioethics, environmental ethics, business ethics, sports ethics, and beyond. In these domains, care, understood as a virtue, or virtues viewed as displays of care, are indeed relevant and promising.

The integration of a plurality of virtues into a unified ethical aim, through Virtue—understood as a general disposition towards the good—is revitalized by this mutual implication and reversibility between care and the traditional virtues. This approach appears both fruitful and promising, as it preserves and expands the classical principle of the unity and reciprocity of virtues.

However, expanding and reconfiguring the virtue ethics framework from the perspective of care, by establishing care as a virtue among virtues, requires addressing a fundamental tension. This tension is inherent to the very notion of virtue: the conflict between 'care for oneself' and 'care for others,' or, in other words, the opposition between its self-regarding nature and foundation, and a other-regarding impulse and vocation. We argue that including care within the system of virtues provides a way to (dis)solve or surpass this dilemma. Accordingly, the following will not focus on the contemporary debate around the recurrent criticism that virtue ethics is inherently self-centered and egoistic (Annas 1995: 127). In fact, taken *simpliciter*, this thesis is simply false: a doctrine of virtues can take both radically egocentric and radically heterocentric forms. We will see this as we proceed.

2 The Metaphysical Grounding of Virtue

We must acknowledge the inherently self-referential nature of virtue. It pertains to an individual's potential, aptitude, and strength in achieving self-fulfillment according to their own nature. The concept of virtue involves the development and perfection of one's capabilities in the pursuit of moral excellence, ultimately leading to *eudaimonia*. This ethical understanding forms the foundation of various forms of moral egoism and is rooted in a metaphysical-ontological framework, as evidenced by the Greek term *aretê* (ἀρετή) and the Latin term *virtus*.

According to Aristotle, "*aretê*" (*ἀρετή*) is an excellence that is ontologically conceived as the actualization (*ἐνέργεια*) of a potential or capacity (δύναμις). In this sense, it implies the achievement of a certain end (*τέλος*) in accordance with the function or work (ἔργον) of a particular nature. Even in Book II of the Nicomachean Ethics, where Aristotle specifically defines the ethical virtues (ἠθικαὶ ἀρεταί) and characterizes the corresponding moral action, he does not contradict this perspective. Before defining ethical virtues (ἠθικαὶ ἀρεταί) as forms of mediacy (μεσότης) between extremes, and to best indicate the type of permanent dispositions or states (ἕξεις)—different from capacities and passions—they are, Aristotle defines them based on the broad sense of excellence (ἀρετή) that applies to all beings and is ontologically grounded:

> [...] every virtue or excellence (*ἀρετή*) both brings into good condition the thing of which it is the excellence and makes the work (ἔργον) of that thing be done well; e.g. the excellence of the eye makes both the eye and its work good; for it is by the excellence of the eye that we see well. Similarly the excellence of the horse makes a horse both good in itself and good at running and at carrying its rider and at awaiting the attack of the enemy. Therefore, if this is true in every case, the virtue of man also will be the state of character which makes a man good and which makes him do his own work well" (Aristotle 1998: *NE* 1106a15-20).

The idea of the full realization of a being is also reflected in the Latin term *virtus*. This is the case, for example, in Cicero, when he says that "virtue (*virtus*) is nothing else than nature perfected and taken to its highest level" (Cicero 1999: 114, *De legibus* 1, 8, 25), even though, of course, it is defined as the conformity of man with his rational nature *(perfectio rationis*).[1]

This notion persisted in the Christian world, namely in the works of St. Thomas Aquinas. In the *Summa Theologica* we read, of course, that a virtue "is a habit of action'" (*habitus operativus*), "good and productive of good" (*boni operativus*), but also that a virtue can be called the "perfection of a power" (*potentiae perfectio*) both in being (*ad esse*) and in acting (*ad agere*) (Aquinas 2020: *Summ. Theol.* Ia IIa, Q. 55, 2, r.). This means that virtue is the ultimate of what a human person can be and fulfil (*ultimum potentiae*) (Aquinas 2020: *Summ. Theol.* Ia IIa, Q. 55, Art. 1. r), the "extreme possibility," and the highest aspiration of man in the full realization of its natural and supernatural vocation (Pieper 2011: 11).

It was perhaps Spinoza that most radically expressed this concept of virtue as the fulfilment of an ontologically inherent potential. The *virtus* is nothing other than the very *"essential"* as *"potential"* of man, that is, *"potestas efficiend"*', in accordance with the VIIIth Definition of the IVth part of the Ethics. Therefore, it can be demonstrated that:

> Virtue is human power itself, which is defined by the essence of a human being alone [...]; i.e. [...] it is defined solely by the endeavor by which a person endeavors to persevere in his own being. Therefore, the more each person endeavors to preserve his own being and is able to do so, the more he is endowed with virtue, and consequently [...], insofar as anyone neglects to preserve his own being, to that extent he is powerless (Spinoza 2018: 174).

[1] When discussing the ultimate goals of man, Cicero inquires. "What is the character of human beings that nature left incomplete? [...] What is it that it must polish and perfect? If there is nothing to be perfected except a certain operation of the mind, namely reason, then the ultimate good must be to live in accordance with virtue. Virtue, after all, is the perfection of reason" (Cicero 2004: 101, *De finibus* IV, 13, 35).

So, in accordance with Proposition 22, virtue manifests the "*Conatus*" as the "strive to persevere in one's own being", and therefore "no virtue can be conceived prior to this one" (Spinoza 2018: 175).[2]

In a very different context and earlier in time, Machiavelli's conception also converges with this notion of virtue as pure and essential power. In fact, virtue acquires a particular meaning and central role in the Florentine's thought, particularly within the conceptual pair of *virtú* and *fortuna* (Machiavelli 2005: 7, 21, 28). The term *virtú* refers, in full conformity with its etymology, to virile force and courage. It also encompasses the capacity for decision and adaptation (*industria*) as forms of actively dealing with *fortuna*—the uncertain flow of events characterized by a mixture of necessity and indeterminism, luck, or fate. Against *fortuna*, *virtú* opposes the wisdom of decision-making (*prudenza*) and the impetus of action (*impeto* and *furore*).[3]

The distinguishing feature of Machiavellian *virtú*, which sets it apart from the Latin *virtus* or the Greek *aretê*, is notable. While the latter are inscribed in a broader morality, Machiavellian *virtú* involves the relativization or transvaluation of ethical-moral qualities and principles that informed or founded the virtuous disposition in ancient or medieval thought, whether it was wisdom, eudaimonia, self-care, or *caritas*.

In the *virtú* that Machiavelli repeatedly praises in *The Prince* and *Discourses on Livy's First Decade*, human inventiveness is combined with animal ferocity. This combination is where its energy and excess lie. In fact, it is a virtue that disassociates from morality and inscribes itself in the realm of the pure exercise of political power and the movement of transforming history. Unencumbered by morality or even openly immoral, *virtú* is essentially the power and excellence in political endeavors. It is an attribute of princes:

> [...] it is not necessary for a prince to possess all of the [...] qualities, but it is very necessary for him to appear to possess them. Furthermore, I shall dare to assert this: that having them and always observing them is harmful [...]. [...] a prince [...] cannot observe all those things for which men are considered good, because in order to maintain the state he must often act against his faith, against charity, against humanity, and against religion. And so it is necessary that he should have a mind ready to turn itself according to the way the winds of Fortune and the changing circumstances command him. And, as I said above, he should not depart from the good if it is possible to do so, but he should know how to enter into evil when forced by necessity (Machiavelli 2005: 61).

The passage unambiguously illustrates the ethical and moral transvaluation of the notion of virtue, while simultaneously staying true to its etymological roots.

---

[2] Quoting Proposition 22: "No virtue can be conceived prior to this one (i. e. the endeavour to preserve itself). Proof: The endeavour to preserve itself is the very essence of a thing [...]. If therefore any virtue could be conceived prior to this virtue, i.e. this endeavour, the very essence of a thing [...] would be conceived as prior to the thing itself. And this (self-evidently) is absurd" (Spinoza 2018: 175).

[3] On the relation and furore, see, for example: Machiavelli (2018: 674–676).

Nietzsche's usage of the concept of *Tugend* is rooted in this dual heritage of *virtus* and *virtù*.[4] Nietzsche refers in various instances to a "moraline-free virtue" (Nietzsche 2005: 4, 6, 85) meaning the power of individual original affirmation, and an impetus for ethical transvaluation. Machiavelli's influence on Nietzsche's thought in this context is direct and recurrent:

> Machiavellianism, *pur, sans mélange, cru, vert, dans toute force, dans tout son âpreté,*[5] is superhuman, divine, transcendent, it is never reached by humans, at most grazed... (Nietzsche 1999: 25–26).[6]

Asserting a virtue free from the morality of the "herd man,"[7] Nietzsche's conception of "the just and the beautiful" is similar in many ways to what Callicles outrageously defends before Socrates in Plato's *Gorgias* (491d-492a). This involves the strength of energy and intelligence in service of passions:

> I tell you now quite frankly that the beautiful and just according to nature is this, that he who would live rightly should allow his desires to become as great as possible and not restrain them, and when they have waxed great, he must be adequate to minister to them with courage and wisdom [...]. But this I think is impossible for most men, whence it is that out of shame they find fault with such people, hiding their own powerlessness [...]. This is why men of better nature are enslaved, as I said before, and those who cannot provide for their pleasures to the full praise temperance and justice due to their own unmanliness (Plato 1984: 279).

By adopting Pindar's injunction "Become who you are!" (Nietzsche 2001: 153, 189) as his own, Nietzsche advances a true ethical-aesthetic construction of character animated by an aristocratic ideal that celebrates strength, creativity, and self-overcoming. He rejects and demolishes virtues tainted by Christianity's "moralin," such as humility, piety, obedience, and chastity (Nietzsche 2002: 86–87), advocating instead for the emergence of alternative virtues better suited to the unique flourishing and achievements of individuals. Ambition, acuteness, and industry are

[4] Nietzsche's position in the debate around "virtue" is paradoxical. On one hand, he is regarded as one of its detractors and critics, and as such, MacIntyre poses the question: "Nietzsche or Aristotle?" (Macintyre 2007: 117). The man of the "great style" who is driven by the "will to power" does not recognize himself in a Kantian moral system and subverts all the traditional and communal tables of virtues. According to MacIntyre, Nietzsche's "moral solipsism" is merely an expression of the "conceptual scheme of individualistic liberal modernity" (Macintyre 2007: 259), to which the Aristotelian tradition provides a fruitful alternative. On the other hand, some American ethicists and commentators have found elements of a "virtue ethics" in Nietzsche's work. On this topic, see: Harcourt (2015: 165–179) and Slote (1998: 23–27).

[5] In French in the original text

[6] The passage in question is from the *Unpublished Fragments* (KSA 13: *Nachgelassene Fragmente* – 1887,11[54](320) 32–35). In these same Fragments are notes of a draft treatise: "about politics of virtue, its means and its ways of attaining power". (fr. 320). This politics of virtue is nothing more than Machiavellianism.

[7] See §199 of *Beyond Good and Evil*: "[...] the herd man of today's Europe gives himself the appearance of being the only permissible type of man and glorifies those morals characteristics that make him tame, easy-going and useful to the herd as the true human virtues, namely: public spirit, goodwill, consideration, industry, moderation, modesty, clemency, and pity" (Nietzsche 2002: 86–87).

qualities to be harnessed against values imposed for common utility, transcending behaviors conditioned by automatism and self-denial, and subverting the narcotizing logic of punishment and reward (Nietzsche 2001: 35).

According to Nietzsche, there is a mature and noble form of egoism, which is a valuable selfishness measured by the value of the self. This depends on whether individuals represent "the ascending or the descending line of life."[8] In this perspective, what truly matters are the character traits and qualities that are founded upon self-love, as they enable self-assertion, the liberation of creative energies, the expansion of one's own power, and the overcoming of oneself. These ideas are all related to the "will to power", which can be seen as an "essentially aretaic (excellence-related) notion" (Swanton 2014: 109).[9] However, in a Nietzschean context, the concept of virtue is not conditioned by any reference to a predefined nature or essence of humanity, nor to any universalizable definition of what constitutes human good or a preconceived idea of moral progress.[10] This is why Christine Swanton characterizes it as "a virtue ethics of becoming" (Swanton: 2015: 200). The Nietzschean view of virtue entails a constant process of self-overcoming towards permanent improvement, but tailored to one's strengths and life situation. Virtue is viewed as a strength and excellence in exercise, and the qualities of character that enable or promote this self-creative becoming are eligible to be considered as "virtues" (Swanton 2015: 199–205).

These brief and concise indications suggest that the conventional notion of "virtue"—as an excellence that fully actualizes one's potential, combining an ontological notion of plenitude with the idea of moral strength for self-realization—may lead to a metaphysical perspective that promotes exalted, egocentric, domineering, and violent subjectivity. Such a perspective could erode openness to diversity and lead to various forms of ethical and moral "transvaluation," or even immoralism.

3 Self and Other Centredness in Virtue Ethics

Given the power for self-assertion, self-promotion, and self-realization that embodies the metaphysical sense of the idea of virtue, it becomes apparent how self-love, self-esteem, self-care assumes primacy in a strictly ethical and moral sense. As Jankélévitch use to assert about Aristotle, "altruism is an irradiation of *philautia* [self-esteem]" (Jankélévitch 1968a: 83), whereby friendship, love, or the broader benevolence can only be possible inasmuch as their object is "another self" (*heteros*

[8] Cited by Douglas Burnham (2015: 114).

[9] The author goes further: "Undistorted will to power plays the same role in Nietzsche's conception of virtue as phronēsis plays in Aristotle's. Understanding and defending this idea will not only enable us to understand the virtues of the mature individual, the 'virtuous egoist', but also enable us to see how Nietzsche can be regarded a virtue ethicist" (Swanton 2014: 109). See also Swanton (2015: 109–179).

[10] In this regard, see §6 of "Morality as anti-nature", from *Twilight of the Idols* (Nietzsche 2005: 175). See also: Swanton (2005: 187, 189–192).

autos) (Aristotle. 1998: 178). Let's recall a famous passage about friendship (1165b5-11):

> For men say that one ought to love best one's best friend, and a man's best friend is one who wishes well to the object of his wish for his sake, even if no one is to know of it; and these attributes are found most of all in a man's attitude towards himself, and so are all the other attributes by which a friend is defined; for, as we have said, it is from this relation that all the characteristics of friendship have extended to our neighbours. All the proverbs, too, agree with this, e.g. 'A single soul', and 'What friends have is common property,' and 'Friendship is equality,' and 'The knee is closer than the shin'; for all these marks will be found most in a man's relation to himself; he is his own best friend and therefore ought to love himself best (Aristotle 1998: 173–174).

For the Stagirite, it is clear that individuals should love themselves, as they are the primary and most immediate objects of their own perception and action. Indeed, the entire system of social and political organizations is based on the assumption that people love and must love themselves.

Of course, this fundamental self-love should not be mistaken for primal selfishness or greed regarding the distribution of goods, honors, and pleasures. Rather, it is a self-esteem rooted in the accomplishment of good actions and the attainment of a noble character. The highest expression of this self-love is the appreciation and admiration of the best parts of oneself, specifically, one's intellectual activity and moral excellence, which represents the ultimate realization of one's humanity. Therefore, it is the very best part of the self that ought to be cherished. Virtue aims at the highest realization of the human potential, as that which is capable of being loved by oneself and admired by others. As Aristotle posits, "the good man should be a lover of self (for he will both himself profit by doing noble acts, and will benefit his fellows)" (Aristotle 1998: 175, 1169a11).

The notion of "*epimeleia heautoû*" (*ἐπιμέλεια ἑαυτοῦ*), as "care of the self", associated with the therapy of passions, the process of conversion and surpassing of the self, in Socrates, Plato and Seneca, is another characteristic expression of this self-referentiality and self-centredness of ancient virtue ethics. Pierre Hadot's well-known works have helped shed light on this philosophical approach to life, characterized by three essential aspects: peace of mind (*ataraxia*), inner freedom (*autarkeia*), and cosmic consciousness (*megalopsychia*). This approach involves a transformative expansion of the self, realizing the greatness of the soul in its belonging to the Cosmos (Hadot 2020: 230). Caring for oneself and striving for excellence, implies both an existential therapy and an aesthetic commitment associated with the concepts of "culture of the self" and "sculpture of the self". Michel Foucault interpreted and adopted these concepts as an "aesthetics of existence" (Foucault 1986: 471–472, 939).

Despite the historical foundation and expressions of this self-centredness, it is evident that the notion of virtue also acquires a strong other-centred character. The responsibility for introducing this dimension to the doctrine of virtues lies with the morality of the Christian world, inspired by the spirit of the New Testament. But here we encounter another challenge. It has indeed been argued that Christian morality, by grounding itself in Divine Law—its commandments and duties—would be inherently at odds with a virtue ethics. As J. B. Schneewind aptly recalls, Thomas Reid had already defended this notion, asserting that the core of Christian morality would not be framed in terms of the classical problem of virtues, but rather "under the three heads of the duty we owe to God, to ourselves, and to our neighbor" (Schneewind 1990: 44). However, such a view seems to overlook the extensive debate on the definition and classification of virtues, which, since the 12th century—starting with Hugh of Saint Victor, Peter Lombard, and Abelard—has continued to intensify in two main traditions, Augustinian and Aristotelian, culminating in a decisive synthesis in the work of Saint Thomas Aquinas (Lottin 1949).

Saint Augustine is a pivotal figure in the transition from the Ancient to the Christian world. His influence was decisive for medieval moral thought even before Saint Thomas re-elaborated Aristotelianism, fully integrating it with New Testament doctrine. This integration particularly combined the cardinal virtues of prudence, temperance, fortitude, and justice with the Pauline virtues of faith, hope, and charity, known as "theological" virtues (Lottin 1949: 66).[11]

The Augustinian approach and tradition to virtue theory introduced the triad of *voluntas* (will), *caritas* (love or charity), and *gratia* (grace), which in some sense steers away from the classical trilogy of *hexis* (permanent disposition), *phronesis* (practical wisdom), and *eudaimonia* (happiness). According to this perspective, virtue depends on the "good will" *(buona voluntas)*, which is "the desire to live with righteousness and honesty" and is essentially moved by love of God and neighbor (*caritas* and *dilectio proximi*). Therefore, the cardinal virtues (prudence, fortitude, temperance, and justice) are always described in Augustine's works as expressions of the good will (Augustine 2010: *De Lib. Arb.* I, 13, 27), love of neighbor (Augustin 2008: *Epist.* 155,16), and love of God (Augustine 2016: *De Moribus*. 15, 25). In a crucial passage of *On the Morals of the Catholic Church*, we find the most concise formulation of this idea of the union of virtues in love, or the understanding of virtues as figures or manifestations of love:

> [...] the fourfold division of virtue I regard as taken from four forms of love. For these four virtues [...], I should have no hesitation in defining them: that temperance is love giving itself entirely to that which is loved; fortitude is love readily bearing all things for the sake of the loved object; justice is love serving only the loved object, and therefore ruling rightly; prudence is love distinguishing with sagacity between what hinders it and what helps it (Augustine 2016: 3902, *De Mor*. 15, 25).

To possess full charity is to possess all virtues, because it implies, assumes, and mobilizes them all. Charity is to love one's neighbor in God and to be directed to Him with all one's heart, soul, and mind. Hence, the famous definition from Book XV of *The City of God* follows: Virtue is "the order of love", or "rightly ordered love" *(ordo amoris)* (Augustine 1998: 680).

---

[11] Odon Lottin explains how the beginnings of treatises on virtues in the Middle Ages were very modest. But, Hugo of St. Victor and Abelard began to seek a general definition of virtue and criteria for dividing virtues. From the beginning, two parallel currents emerged: with Hugo of St. Victor, Peter Lombard, and Peter of Poitiers, a current of ideas quite Augustinian; with Abelard, on the contrary, and Simon of Tournai, an Aristotelian current. (Lottin 1949: 99).

Such an ethical and moral openness towards alterity, both the contingent otherness of neighbour, and the absolute otherness of God, is essential for developing an other-regarding virtue theory. This morality is driven by a will imbued with love as a virtue, which is gracious giving, and sacrifice, which focuses on the good of the other.

Besides, it is worth noting that Augustine's doctrine of original sin brought about a significant shift in the conception of human beings. The ancient world's moral perfectionism relied on an optimistic view of human nature and will. On the contrary, Augustine presented a predominantly negative view of self-love and a deeply corrupt understanding of the postlapsarian human nature. *Amor sui*, or self-love, is portrayed as the primary reason for humanity's fall into rebellion against God. It is the dominant passion, the source of all vices, and the root of humanity's sins, leading to greed, pride, and impiety (Maurer 2019: 18–19). Thus, while essential, the will alone is insufficient to attain virtue. It must be accompanied by Grace. What is significant is that Augustinian moral agency embraces a fundamental and radical Otherness, referring to what is "more inward than my innermost self, higher than my highest" (Confessions III, 6, 11).

Let us now briefly consider, from an aretaic perspective, the fundamental biblical commandment of love: "You shall love your neighbor as yourself," (*Leviticus* 19:18, *Matthew* 22:39, Mark 12:31, 33). It is the "as yourself" that is problematical. It cannot mean loving the other as a mere analogical projection of natural self-love. On the contrary, as Jankélévitch tells us, "it is the Other who serves as the self; the lover is a self who has its self outside itself," or, in other words, "the Thou is the core of the self" (Jankélévitch 1968b: 890–891).

We are, therefore, at the antipodes of the principle of the alter ego and Aristotle's philautia: "It is not in an Aristotelian sense, but in an ecstatic and truly miraculous sense that Leviticus commands us: Love your neighbour as yourself; that is to say, not: love your friend as an appendage of yourself, nor: love yourself in your friend; but rather: You will love your neighbour as much as he loves himself; the self loves as if it, the lover, did not exist, the self has no other 'self' than its beloved, the self becomes its 'other' in person through the force of love; the loving self is entirely emptied of its egoity." (Jankélévitch 1968b: 799–800). Such a superlative commandment can only be lived through a radical conversion of the will, within a virtue that reconfigures the entirety of agency. The forgetting of oneself in ecstatic coincidence with the other is the limit of a "pure love of charity," which finds its most radical expression in Fénelon.

This interpretation of the evangelical law of love allows us, bypassing Modernity and moving into the twentieth century, to consider one of the most original theories of virtue in contemporary thought: that of Vladimir Jankélévitch's *Traité des Vertus* (1949).

His conception of virtue as love (*amour* or *charité*), imbued with a fundamental heterocentric vocation and in line with the Augustinian tradition, is notable for its de-ontologized approach to the aretaic theory. Virtue, as a disposition of the will, quality of intention or force of acting in action, consists in aiming at and achieving the good of the other, in an absolute and selfless way. Jankélévitch could not be clearer in his rejection of the idea that "ontic positivity" could coincide with "normative plenitude" through the union of virtue with the power of self-assertion. In fact, the "moral vocation" and the "ethical imperative" invalidate "the great physical, elementary, and primary evidence of *conatus*" (Jankélévitch 1968a: 13). In the domain of ethics and virtue, and such is the thesis, "the preservation of *self*-being and the increase of *self*-having come, henceforth, last" (Jankélévitch 1981: 167). The "desire to be" is replaced by the "desire to be for the other and to make the other's being be" (Jankélévitch 1968a: 13). The true "original affirmation"[12] consists in the affirmation of the other's being, so that renouncing the spontaneous and natural centrality of the ego and promoting a creative self-abnegation or denial are the starting point and the radiating center of virtue. Such is the radical heterocentrism in which the moral subjects seem to even renounce, in themselves, the universal dignity of the human being. Such a subject declares, therefore, "everyone has rights, except me!" (Jankélévitch 1968a: 10). Such a stance consists in affirming that the "ontic minimum" is the condition of the "ethical maximum" (Jankélévitch 1981: 118).

4 Self-Regarding Virtues and Other-Regarding Virtues

The paradoxical tension present in virtue theory between self-care and care for others is reflected, explicitly and implicitly, in the contemporary common distinction between self-regarding virtues and other-regarding virtues. It is interesting to note that this distinction was first outlined by Hume, Bentham, Mill, and Sidgwick,[13] before being clarified by authors such as G. H. Von Wright (1963: 153), Gabriele Taylor, and Sybil Wolfram (1968: 238–248), and becoming a leitmotif of the "virtue ethics" movement in authors such as M. Slote, M. Stocker, Ph. Foot or David Carr.

Self-regarding virtues allegedly promote the interests and well-being of the agent who possesses and practices them. So, they can be call self-referential, intrapersonal, ascetic, self-control, or prudential virtues. *Other-regarding virtues* typically benefit and promotes the well-being of others, whether collective or individually. They are often call other-referential virtues, interpersonal or social virtues. This distinction is employed by the forementioned authors (e.g., Von Wright 1963; Slote 1995; Foot 2001) who list multiple and diverse virtues responding to this characterization. Fortitude, and perseverance are examples of self-regarding virtues. Benevolence, generosity, and justice are examples of the other-regarding ones. Consensually, it could be said that temperance is the paradigm of the virtue centered

[12] In the sense of Jean Nabert, which means a "desire to be", an effort to exist, to which we are always unequal (Cf. Nabert 1992: 125, 130, 138). The "moral effort" is nothing more than what the subject affirms by his initiation of that absolute affirmation that inhabits him. The "moral imperatives" are "a condition of the development of our desire to be" (Nabert 1992: 138 and 139). The opposition with Jankélévitch could not be more clamorous.

[13] For an overview on this matter, see: Albee (1897) and Llobera Trias (2014).

on the agent's own good, and justice the prototype of the virtue centered on the good of others.

Given the difficulty of a clear distinction and classification of traditional virtues such as courage, according to this criterion, Michael Slote suggests the existence of *mixed virtues* (Slote 1995: 92–108) that both benefit the moral agent and promote the good of others. Acts of courage carried out by courageous men can serve the interest of the individual as well as the good of other persons or the community. It also happens that self-referential virtues can, in a circumstantial or accidental way, directly or indirectly benefit others. Temperance or frugality, as well as an individual's industrious character, benefit the community around him, even though they are not to be confused with generosity or altruistic zeal.

But below and beyond these difficulties of mutually exclusive dual classification, inherent to the intimate congeniality or consubstantiality of virtues in virtue, this dichotomy shows an asymmetry between the attention paid to the self and to others in ethical theories, in general. This leads to an oscillation between perspectives understood as *egoistic* and *altruistic* regarding the agent of virtues. Michael Slote refers to these as "agent-favoring self-other asymmetry" and "agent-sacrificing self-other asymmetry" (Slote 1995: 6; Slote 1984). The first kind holds that it is morally admitted to favor our own interests over those of others, while the second holds that it is morally admitted to sacrifice our own interests for others.

Meanwhile, David Carr introduces the distinction between *virtues of self-control* and *virtues of attachment* that adds complexity to the debate and challenges the initial self-referential and other-referential criteria. These two classifications are not mutually exclusive; instead, they overlap and intersect with each other, resulting in a more nuanced understanding of the concept of virtue:

> [...] that class of character traits I wish to distinguish from the virtues of self-control as the virtues of attachment includes such other-regarding attitudes as unselfishness, considerateness, sympathy, benevolence, kindness, generosity, courtesy, respect, charity and possibly patience and tolerance, as well as such self-regarding virtues as modesty and humility (Carr 1985: 53).

According to David Carr, it is acceptable to label self-regarding virtues such as modesty and humility as *virtues of attachment* because they imply a serious consideration of one's own worth in relation to others, as well as a serious consideration of other people.

Despite these classification difficulties or inconsistencies, there has been an ongoing debate about the primacy of self-regarding aspect of virtue theory, as defended by Foot (2001: 78–79), versus the priority of other-regarding aspect of virtue theory, as defended by authors like Goldin (2001) and Bloomfield (2011). However, some authors such as Carr (1985) and Slote (1999) propose a mixed conception of virtue ethics that seeks to balance the other-regarding and self-regarding aspects of aretaic morality. In the end, the question is about which sphere of normativity should be seen as fundamental, not only in virtue theory but in moral theory in general: should it be the primacy of the normative sphere of the relation with oneself or the normative sphere of the relation with others?

We believe it's essential to overcome the chiasmus of asymmetries between the self and others in virtue theory. Neither other-regarding virtues should be thought as derived solely from self-regarding virtues, nor should the opposite be true. Such a reductionist understanding of the relationship between self and others is inadequate. When reading these authors, one can observe that the distinctions and classifications they propose often lead to a maze of difficulties. This is because the criteria of distinction, whether characterologically or casuistically applied, tends to overlook the fundamental fact that the different virtues are mutually implicated and solicited in their exercise and the pursuit of the good. Therefore, it is not sufficient to simply say that both kinds of virtues we have considered are important for human well-being, since both our own interests and those of other people are crucial for a "decent human life", as some authors have suggested. Instead, we should consistently add and emphasize, as Jankélévitch does, that unlike duties and norms, instincts and pleasures, virtues do not compete with each other, but rather require each other reciprocally, and can even transform into one another (Jankélévitch 1968a: 307–309).

In fact, a fundamental and unified virtuous intention, aiming at the accomplishment of good, here and now, lies within any virtuous trait or character expressing and exercising itself concretely. Let's give some short examples. How is it possible to be just and generous in a situation of financial crisis or on a war context without being courageous and temperate? Is the bravery or sagacity shown by a thief who violates even the most basic principles of justice praiseworthy as virtues? Are the truthfulness of a delator or sincerity that serves cruelty, virtuous? The unity of the plurality and diversity of *virtues* lies in *Virtue* itself, which is a benevolent intention and a beneficial action, both expressions of a virtuous disposition actively oriented towards the Good.

5 Surpassing the Dilemma

It is possible to overcome the difficulties posed by characterological and casuistic distinctions, avoid the symmetrically reductionist choice between an egoistic and a sacrificial perspective, and move beyond an opposition that reinforces a profound asymmetry between *self* and *other* in moral action and ethical life. To achieve this, we must accept the following five principles or assumptions:

1. The principle of organic unity, mutual implication, and reciprocity of virtues, which means reconducting or reconciling the plurality of *virtues* with the unity of ethical aims in *Virtue*.
2. The intrinsically relational dimension of moral action, integrating both hetero and self-regarding motivations and values.
3. The intersubjective configuration of moral subjectivity and agency as a "relational self."
4. The recognition of care as a virtue among virtues, transforming the all aretaic doctrine.

5. The re-evaluation of the idea of perfection, and the notion of moral perfectionism based on the previous assumptions and within a dialogical rather than an egological dynamic.

Only this dialogical principle and mutual implication can prevent the interminable casuistic analyses where considerations between self-regarding virtues, other-regarding virtues, and mixed virtues linger and get bogged down.

Including care in the board of virtues is, in fact, a pivotal strategic point. It helps to overcome this artificial duality between self *versus* other centredness or referentiality, for caring for oneself is a necessary condition for caring for others. Actually, it is important to note that care theorists reject a self-sacrificial conception of care (Gilligan 1983: 74), just as virtue ethicists reject moral egoism (Annas 1995: 127). Since we are defined in relation, "caring is, thus, both self-serving and other-serving" (Noddings 2013: 99). An ethics of virtue is therefore at most formally self-centered or egoistic, for "its content can be fully as other-regarding as that of other systems of ethics" (Annas 1995: 127).

In fact, and most generally, self-regarding dispositions and skills enable other-regarding abilities: there is no justice or sincerity without courage or prudence, just as there is no generosity without temperance, nor benevolence without magnanimity and humility. It is also clear that promoting the good of others requires not only other-regarding virtues and motivations but also eliciting and encouraging the acquisition and development of self-regarding virtues in others. Since virtues are *admirable character traits* and abilities developed through practice, promoting them involves a convergence of self-care and care for others within a system of virtues that is an open totality, with each part expressing the whole.

The foregoing discussion prompts a revaluation of the concept of "moral perfectionism," which must not be conflated with self-centeredness or moral egotism. Instead, the pursuit of moral excellence should be understood in relation to the overarching ideal of goodness. In this context, Iris Murdoch's philosophy offers a valuable perspective by emphasizing the importance of moral development in relation to the "ideal of perfection" and by challenging us to transcend mere egological self-assertion. She recognizes the interdependence between moral endeavor and the common good in the cultivation of virtuous character.

According to Murdoch, such perfectionism entails a mode of moral progress that pertains to the totality of the person in reference to the Good. The Good, which is both inexhaustible and indefinable, embodies the "non-metaphysical meaning of the idea of transcendence" (Murdoch 2014: 91). This transcendence is experienced through immanence as a process of continual inner transformation and self-surpassing. Hence, the fundamental ethical-moral inquiry becomes, not just "how should one live?", but "How can we make ourselves better?" (Murdoch 2014: 81). This question comes from below and goes far beyond establishing a formal criterion for an isolated decision or a single action and instead signifies a persistent aspiration that extends to all facets of life. In fact, moral action is the product of the "texture of a man's being", that is of the "nature of his personal vision" (Murdoch 1999: 97–99). His whole character and consciousness are mobilized and involved on a permanent transformation and governed by an "idea of perfection", wish refers to the Good as a "sovereign concept" (Murdoch 2014: 23). The primary condition for perfectibility is the overcoming of selfish impulses and the expansion of consciousness to the otherness and distance of the world and persons. As Iris Murdochs asserts: "To say that the good is a transcendent reality means that virtue is the attempt to pierce the veil of selfish consciousness" (Murdoch 2014: 91). Or, as she also argues: "The chief enemy of excellence in morality (and also in art) is personal fantasy: the tissue of self-aggrandising and consoling wishes and dreams which prevents one from seeing what is there outside oneself" (Murdoch 2014: 57).

Virtue is concerned with truly apprehending that other people exist. In the face of the otherness and plurality of the world and its people, confronted with the disjointed character of values that do not harmonize within any rational axiology, but rather tear apart the firmament of morality in conflict, only the reference to the Good through love provides a form of unity and indicates an ascendant path. Virtues are manifestations of goodness that belong to the "continuous fabric of being" (Murdoch 2014: 21–22). Therefore, only the love inner to all virtues can cross that tension towards the Good, a love that animates all virtues.

Although not being virtue ethicist *stricto sensu*, Iris Murdoch defends the thesis of the unity and convergence of virtues:

> If we reflect upon the nature of the virtues we are constantly led to consider their relation to each other. The idea of an 'order' of virtues suggests itself, although it might of course be difficult to state this in any systematic form. For instance, if we reflect upon courage and ask why we think it to be a virtue, what kind of courage is the highest, what distinguishes courage from rashness, ferocity, self-assertion, and so on, we are bound, in our explanation, to use the names of other virtues. The best kind of courage (that which would make a man act unselfishly in a concentration camp) is steadfast, calm, temperate, intelligent, loving. [...] Whether there is a single supreme principle in the united world of the virtues, and whether the name of that principle is love, is something which I shall discuss below. All I suggest here is that reflection rightly tends to unify the moral world, and that increasing moral sophistication reveals increasing unity (Murdoch 2014: 56).

Often, Iris Murdoch takes up the idea, already so dear to the Stoics, that virtues are not possessed separately. She assumes the Augustinian belief that all virtues are expressions of love and goodwill, and also adopts the Aristotelian principle that all are guided by a form of practical wisdom. Asserting that the unity of virtues is governed by the Good, beyond any egocentric confinement of moral conscience and agency, Murdoch points us to the path.

6 Conclusion: The Virtue of Care Between Autonomy and Vulnerability

The ethics of care emphasizes the "vulnerable condition" of the human being, highlighting the dependence and fragility of the moral patient amidst the challenges and tensions of the contemporary—whether post-modern or hyper-modern—world.

However, in this same context, it is crucial to acknowledge that the moral agent is both capable and fallible, as Ricœur reminds us (Ricœur 2005), so it is necessary to understand that caring for oneself and caring for others are intertwined and mutually dependent.

Thus, the ethics of autonomy and the ethics of vulnerability are reunited in a virtue theory transformed through the incorporation of care as a virtue, and implicated in a reconfiguration of the notions of "goodness" and "moral perfectionism." The development of ethical-moral dispositions, aptitudes, and skills that we can designate as virtues implies the convergence of caring for oneself and caring for others. Without the development of the self-regarding virtues of courage, temperance, practical wisdom, and humility, the moral agent will not be effectively able to exercise the other-regarding virtues of justice, generosity, and sincerity.

The foundational questions of ethics and moral life—'How should one live?' and 'How can we make ourselves better?'—do not establish a insurmountable division and insurmountable opposition and division between self-referential and other-referential motivations. The normative spheres of one's relationship with others and with oneself are intrinsically interconnected. Thus, conceiving, enacting, and cultivating a virtue of care serves to strengthen this interconnected perspective.

References

Albee, Ernest. 1897. Hume's ethical system. *The Philosophical Review* 6 (4): 337–355.
Annas, Julia. 1995. *The morality of happiness*. New York: Oxford University Press.
Aquinas, Thomas. 2020. The English translation of the *Summa Theologiae*. Trans. Fr. Laurence Shapcote, revised by The Aquinas Institute. https://aquinas.cc/la/en/~ST.I-II.Q55.A2. Accessed 23 April 2023.
Aristotle. 1998. *Nicomachean ethics*. Trans. D. Ross. Oxford: Oxford University Press.
Augustin. 2008. *Letters: Volume III (131–164)*. Trans. Wilfrid Parsons. Washington: The Catholic University of America Press.
Augustine. 1998. *The city of god against the pagans*. Trans. R- W- Dyson. Cambridge: Cambridge University Press.
———. 2010. *On the free choice of the will, on grace and free choice, and other writings*. Trans. Peter King. Cambridge: Cambridge University Press.
———. 2016. On the morals of the Catholic Church. In *Collected works of Saint Augustine*. Trans. Richar Stothert (ebook). Hastings: Delphi Classics.
Beato, José. 2021. Care: A virtue among virtues. In *Philosophy of care. New approaches to vulnerability, otherness and therapy*, Advancing global bioethics, ed. Joaquim Braga and Mário Santiago de Carvalho, vol. 16, 195–210. Springer.
Bloomfield, P. 2011. Justice as a self-regarding virtue. *Philosophy and Phenonemological Research* 82 (1): 46–64.
Burnham, Douglas. 2015. *The Nietzsche dictionary*. London: Bloomsbury.
Carr, David. 1985. Two kinds of virtue. *Proceedings of Aristotelian Society New Series* 85: 47–61.
Cicero. 1999. *On the commonwealth and on the Laws*. Trans. James E. G. Zetzel. Cambridge: Cambridge University Press.
———. 2004. *On moral ends*, 2004. Trans. R. Woolf. Cambridge: Cambridge University Press.
Foot, Philippa. 2001. *Natural goodness*. Oxford: Oxford University Press.

Foucault, Michel. 1986. *The care of the self. Volume 3 of the history of sexuality*. Trans. Robert Hurley. New York: Pantheon Books.
Gilligan, Carol. 1983. *In a different voice: Psychological theory and women's development*. Cambridge, MA: Harvard University Press.
Goldin, O. 2001. Porphyry, nature, and community. *History of Philosophy Quarterly* 18 (4): 353–371.
Hadot, Pierre. 2020. *The selected writings of Pierre Hadot: Philosophy as practice (re-inventing philosophy as a way of life)*. London: Bloomsbury Academic.
Harcourt, Edward. 2015. Nietzsche and the virtues. In *The Routledge companion to virtue ethics*, ed. Lorraine Besser-Jones and Michael Slote. New York: Routledge.
Jankélévitch, Vladimir. 1968a. *Traité des vertus 1. Le sérieux de l'intention*. Paris: Bordas-Mouton.
———. 1968b. *Traité des vertus 2: Les vertus et l'amour*. Paris: Bordas-Mouton.
———. 1981. *Le Paradoxe de la morale*. Paris: Seuil.
Llobera Trias, Ignasi. 2014. *Les 'other-regarding virtues' en L'ètica contemporània. Tesi di doctorat*. Barcelona: Universitat de Barcelona.
Lottin, Odon. 1949. *Psychologie et morale aux XIIe et XIIIe siècles. Tome III: Problèmes de morale. Seconde Partie, I*. Louvain: Duculot.
Machiavelli, Niccolò. 2005. *The prince*. Trans. and ed. by Peter Bondanella. Oxford: Oxford University Press.
———. 2018. Discorsi sopra la Prima Deca di Tito Livio. In *Tutte Le Opere: Secondo L'edizione Di Mario Martelli (1971)*. Firenze: Bompiani.
Macintyre, Alasdair. 2007. *After virtue: A study in moral theory*. 3rd ed. Notre Dame: University of Notre Dame Press.
Maurer, Christian. 2019. *Self-love, egoism and the selfish hypothesis: Key debates from eighteenth-century British moral philosophy*. Edinburgh University Press.
Murdoch, Iris. 2014. *The sovereignty of good*. London and New York: Routledge.
———. 1999. *Existentialists and mystics: Writings on philosophy and literature*. New-York/London: Penguin.
Nabert, Jean. 1992. *Eléments pour une éthique*. Paris: Aubier.
Nietzsche, Friedrich. 1999. *Nachgelassene Fragmente 1887–1889 Kritische Studienausgabe 13*, 1999. München: Gruyter.
———. 2001. *The gay science: With a prelude in German rhymes and an appendix of songs*. Trans. Josefine Nauckhoff and Adrian Del Caro. Cambridge: Cambridge University Press.
———. 2002. *Beyond good and evil: Prelude to a philosophy of the future*. Trans. Judith Norman. Cambridge: Cambridge University Press.
———. 2005. *The Anti-Christ, ecce homo, twilight of the idols, and other writings*. Trans. Judith Norman. Cambridge MA: Cambridge University Press.
Noddings, Nel. 2013. *Caring, a feminine approach to ethics & moral education*. Berkeley: University of California Press.
Pieper, Josef. 2011. *The Christian idea of man*. Trans. Dan Farrelly. South Bend: St. Augustine's Press.
Plato. 1984. *Euthyphro, apology, Crito, Meno, Gorgias, Menexenus*. Trans. R. E. Allen. Vol. 1. New Haven: Yale University Press.
Ricœur, Paul. 2005. *The course of recognition*. Trans. David Pellauer: Harvard University Press.
Schneewind, J. B. 1990. The misfortunes of virtue. *Ethics, 101*(1), 42–63.
Slote, Michael. 1984. Morality and self-other asymmetry. *The Journal of Philosophy* 81 (4): 179–192.
———. 1995. *From morality to virtue*. Oxford: Oxford University Press.
———. 1998. Nietzsche and virtue ethics. *International Studies in Philosophy* 30 (3): 23–27.
———. 1999. Self-regarding and other-regarding virtues. In *Virtue ethics and moral education*, ed. David Carr and J.W. Steutel, 95–106. London/New York: Routledge.
Spinoza, Benedict de. 2018. *Ethics proved in geometrical order*. Trans. Michael Silverthorne and Matthew J. Kisner. Cambridge: Cambridge University Press.

Swanton, C. 2005. Nietzschean virtue ethics. In *Virtue ethics, old and new*, ed. Stephen Gardiner, 179–192. Ithaca: Cornell University Press.
———. 2014. Nietzsche's virtue ethics. In *The handbook of virtue ethics*, ed. Stan Van Hooft, 105–117. Durham: Acumen.
———. 2015. *The virtue ethics of Hume and Nietzsche*. Oxford: Wiley Blackwell.
Taylor, G., and S. Wolfram. 1968. The self-regarding and other-regarding virtues. *The Phiosophical Quarterly* 18 (72): 238–248.
Von Wright, G.H. 1963. *The varieties of goodness*. New York: The Humanities Press.

Chapter 6
Dancing on the Edge of the Abyss

Teresa Martinho Toldy

Abstract According to Kristen Dunphy's presentation of the TV series "Wakefield" (2020): "There's a fine line between sanity and madness". Every episode of this series starts with a psych nurse standing at the edge of an abyss in the Australian Blue Mountains. In fact, the psych ward where he works is situated near to that abyss. Sanity and madness are presented as something that goes far beyond a simple clinical situation: they may be considered almost as a symbol of a vulnerability common to each and every human being: we all stand at the edge of an abyss—the abyss of mortality. And yet, at the same time, the notion that, even when you have to take care of someone, you are also hit by your own fragility, results in an interpretation of human existence as a complex web of caring and being cared for. Our common vulnerability may lead human beings to be drawn into the abyss or to dance on its edge, even if that dance is precarious. This text focuses on the awareness of our common vulnerability and precarious condition, as defined by Judith Butler, as grounds to build an ethics of care.

Keywords Care · Madness · Precariousness · Sanity · Vulnerability

1 Introduction

Let us start with the story of a TV series called "Wakefield" (Dunphy 2020). Wakefield is a psychiatric ward situated in the Australian Blue Mountains. Every episode of "Wakefield" starts with a psychiatric nurse (Nik) standing at the edge of an abyss in those mountains, or dreaming about being there. He is a nurse who, according to Kristin Dunphy, the creator of the drama series, "has a gift: a gift for soothing the afflicted and reaching the unreachable" (Dunphy: 4). Nik is the "go-to

T. M. Toldy (✉)
Fernando Pessoa University, Porto, Portugal

CITCEM, University of Porto, Porto, Portugal
e-mail: toldy@ufp.edu.pt

J. Braga (ed.), *Ethics of Care*, Advancing Global Bioethics 20,
https://doi.org/10.1007/978-3-031-73608-7_6

man in emergencies", combining expertise with intuition in moments when no one seems to know how to control violence, despair, isolation, self-denial, addictive behaviours of various types (drugs, sex, gambling) and fears of loneliness. He acts and speaks with patients in full awareness of the condition of each and every one of them and is also aware of the failures existing in that particular psychiatric ward near the abyss in the Australian Blue Mountains. He is mindful of the lives, relationships, families and jobs of all characters, the pride, sense of self-worth or self-unworth that contextualize the various behaviours of the patients. Nik's "special gift" reveals itself in his capacity to connect with others, interacting at the same time as a psychiatric nurse and as a person that sees patients as more than just patients. Nik sees them as people, not merely cases of a specific mental illness.

However, throughout the series, we do not only see pathologies in the patients. The carers (nurses, psychiatrists and bureaucratic staff) are also wounded people. Kristin Dunphy draws our attention to the fact that patients "are being cared for by staff who often have mental health issues of their own" (Dunphy: 8). There is a nurse, insecure of her position as chief-nurse, who is always malicious and manipulative. She has no life outside the clinic, but invents the existence of a dependent daughter as an excuse for her bitterness and inability to feel compassion for patients or build friendships with her colleagues. She does nothing to improve the condition of a young patient in a catatonic state, since this ensures his dependence on her. There is also a doctor with a terrible sense of direction who gets hopelessly lost on a regular basis. A naïve carer who is a single mother of two and just wants to be loved and recognised for her generosity. A young psychiatrist terrified of his patients, drowning in anxiety and despair. A nurse who is indifferent to everything around him, insensitive, morbidly cynical and blatantly inconsiderate.

Nik lives his day-to-day life trying to improve the clinical situation of the patients using simple and discreet psychological techniques combined with a sensibility that enables him to interact with them as human beings, recognizing and appealing to their common ground: the fact that both the clinical staff and the patients are (wounded) human beings. However, there is something mysterious about Nik's life: a specific song keeps coming to his mind, something that he is not able to understand until the day his hidden memories of a tragedy in his family's past hit him, striking him in such a brutal way that Nik, the "go-to man in emergencies", falls into a severe psychiatric crisis. The song he keeps hearing in his mind is the song that was playing on the radio while he was bathing his younger brother, in his childhood. The radio fell into the bathtub and killed his little brother with an electric shock. The remembrance of the past that emerges from Nik's "hidden memories" during a conversation with his sister at her wedding ceremony, has the effect of an avalanche. The series ends with Nik in the miserable situation of some of the ward's patients before the eyes of a horrified staff. After all, what Kristin Dunphy, the creator of this TV series wanted to point out was that "there is a fine line between sanity and madness" (Dunphy: 8).

Sanity and madness are things that go far beyond a simple clinical situation. The aim of mentioning this story is not to speak specifically about this fine line between sanity and madness. Rather, this "fine line" symbolically highlights a condition common to us all. We all teeter on the edge of an abyss—the abyss of not being totally able to cope with situations that hurt us, the abyss of being destitute, dispossessed, the abyss of vulnerability, precariousness and precarity. There are moments in life where experiences of extreme vulnerability cause wounds, wounds that may be reopened when something triggers us, and reminds us of our proximity to the abyss. But those wounds, in fact, merely lay bare the vulnerability that is a feature of the human condition. We could call it the anthropological wound that emerges in situations of illness, pain, sorrow, grief, abandonment, exclusion.

André (2021), in his extremely interesting text about the relationship between an anthropology of Vulnerability and the ethics of care, reminds us that we all suffer from a wound: the wound of vulnerability. And yet, returning to the last episode of "Wakefield", we are confronted with a symbolic scene in which Nik and the catatonic patient, now freed from the hands of the chief-nurse, dance at the edge of the abyss. Putting it bluntly: even when you have to take care of someone, you are also afflicted with your own fragility. The awareness of this vulnerability may be understood in a negative or in a positive sense. It is a wound and, at the same time, it opens the door to an interpretation of human existence as a complex web of caring and being cared for, both as an ontological condition and as a political obligation. Our common vulnerability may lead human beings to be drawn into the abyss or to dance with other(s) on its edge, even if that dance is precarious.

This text will focus precisely on the awareness of our common vulnerability as the ground to build an ethics of care. It will follow the "path" of Judith Butler's notion of "vulnerability", "precariousness, "precarity", "interdependence", and "dispossession" as guides for a reflection on the possibility or the relevance of thinking about the ethics of care. We will see that Butler uses those notions as the key for ethics, since, according to her, "the question of ethics is always a question of an ethical relation" (Butler and Athanasiou 2013: 107) and the core issue of caring. The text will also recall "Wakefield" now and then: sometimes, a good story has the power to "rephrase" traumatic situations and may be deeper than philosophical arguments.

2 A Bond Between an "I" and a "We"

Butler considers that "the 'I' is invariably implicated in the 'we'" (Butler and Athanasiou 2013: 107). Recognizing this bond is to recognize that "care for another" always implies both acknowledging personal vulnerability and the political relevance of resisting "a social condition of inequality" (idem), that is, a precarity that is imposed. This means that Butler's notion of "care for another" encompasses both a human condition and a political engagement in favour of a "good life" for all. It is not "an individual disposition" (idem). She states: "it is precisely because I am from the start implicated in the lives of the other that the 'I' is already social, and must begin its reflection and action from the presumption of a constitutive sociality" (idem).

Understanding the path taken by Judith Butler to outline her key concept of "caring for another" involves an analysis of her concepts of vulnerability, precariousness, precarity, and dispossession, even if sometimes the boundaries between them (especially between precariousness and precarity) seem to be blurred.[1] Butler is clearly not concerned with a notion of "care" as an applied ethics in the field of health, for instance. Her aim is to frame "care" from a philosophical point of view as a web of caring and being cared for, as the key part of what it means to live as a human being, what she also calls "an interdependency of lives that are mutually implicated in one another" (Butler and Athanasiou 2013: 107).

Butler deems it necessary to "struggle for a conception of ethical obligation that is grounded in precarity" (Butler 2015: 119). She develops the notion of the precarity of each and every human being as a "human condition",[2] but this same notion is the one she uses as the ground for an ethical responsibility, since "precarity exposes our sociality, the fragile and necessary dimensions of our interdependency" (idem). Getting back to the Wakefield story: Nik does not dance alone. Dancing with the former catatonic patient also shows the relevance of interconnectivity: the past of this former catatonic patient is not clear. What is clear is that Nik and he dancing on the edge of the abyss is a manifestation of our "social existence as bodily beings who depend upon one another for shelter and sustenance and who, therefore, are at risk of (…) homelessness, and destitution under unjust and unequal political conditions" (Butler 2015: 118).

Going deep into the social meaning of interconnectivity or interdependency according to Butler, our precarity also depends on, is connected with and arises from social and economic relationships and infrastructures. She does not want to rehabilitate humanism. She aims to advocate for a conception of ethical obligation that is grounded in precarity. There is a certain ambiguity in her notion of "precarity", as already mentioned: sometimes it seems that "precariousness" and "precarity" mean the same thing. Other times it is clear that Butler considers "precarity" as something that is "socially and economically induced" (Butler and Athanasiou 2013: 162), while precariousness seems closer to the notion of "vulnerability".

3 Ontological Vulnerability

As we know, Butler's development of the concept of precariousness and of vulnerability has a connection with her reflections on 9/11. In *Precarious Life* (2004) and in *Frames of War* (2009), her main question was: who is grievable?[3] Who deserves to be grieved? Those questions were directly related to the fact that after 9/11 there

[1] See Miller (2020).

[2] That is why Engster's (2019) reading of Butler's notion of vulnerability, circumscribing it to "social institutions and cultural norms", seems to be somewhat superficial or, at least, incomplete.

[3] Term coined by Butler.

seemed to exist a double standard in regard to mourning: the American victims of this unquestionably brutal terror attack seemed to be considered "grievable", while the Afghanis civilians, victims of the American reaction to the attack, were ignored, as if deemed unworthy of being "grievable". Butler poses a decisive question: "what counts as a liveable[4] life and a grievable death?" (Butler 2004: xv). Are vulnerability and precariousness notions that can only be used to describe sporadic situations or are they ontological notions, that is, notions that describe human beings as such? Are we discussing a sand storm or a statue of clay—an unexpected and sporadic event or a piece of work, still fragile, still precarious, but existing in time, even if within a limited timeframe? Butler's main point is that everyone deserves to be grievable, since each and every human being is vulnerable. In that sense, Butler frames vulnerability and precariousness within the imponderable dimensions of our daily life, in sickness and in health, in desperate situations and in creative ideals that enable us to keep hope alive, to go on dancing even if it is on the edge of an abyss.

Butler's notion of vulnerability does not reflect a pessimistic ontology. On the contrary: she challenges us to think about it both as something that appeals to the recognition of the humanity of every human being, and also as an ontological openness to the Other. If we open our eyes to the world in which we are living, we know there is an abyss, that we may be at the edge of an abyss, but we have to try to keep dancing on its edge, not because we are unaware of what is happening, but because vulnerability also represents a chance to recognize that we all share the same condition.

Actually, there is a previous stage to the acknowledgement of this common vulnerability. This acknowledgement already shows a positive answer to the question: "Who counts as human?" (Butler 2004: 20). The answer to this question is connected with shared experiences of loss: everybody lost someone. Everybody has suffered, in one way or another, the experience of the vulnerability of our bodies (in grief, when menaced by something that was/is far beyond our possibility of control). Therefore, we experience a vulnerability that is ontological, but also social and embodied. We are "socially constituted bodies, attached to others, at risk of losing those attachments, exposed to others, at risk of violence by virtue of that exposure" (Butler 2004: 21).

However, this "exposure to others" (Butler 2004: 21) (in our mourning expressions resulting from the loss of someone we love) produces a change, a transformation over time in a way that may be still unknown and that will reveal itself in the future. It may express itself in the question: "who 'am' I, without you?" (Butler 2004: 22). Even if we think that this question, and the pain attached to it, is private, actually, according to Butler, it is "political", that is, it "furnishes a sense of political community (…) bringing to the fore the relational ties that have implications for theorizing fundamental dependency and ethical responsibility" (idem). Grief is, therefore, a kind of string that connects us, that is, a relational thread: "all of us have some notion of what it is to have lost somebody. Loss has made a tenuous 'we' of us all" (idem).

[4]Term coined by Butler.

4 A Web of Interdependencies

This connection means that we live in a web of interdependencies: the recognition of vulnerability when facing others who are also vulnerable opens the possibility of relationality, since vulnerability is not only a specific situation; rather, it is a constitutive element of human beings living in the world. Butler also thinks we can consider it "an ongoing normative dimension of our social and political lives" (Butler 2004: 27). It is at the root of ethics and it should be at the root of the political organization of societies: we are all vulnerable bodies. However, our vulnerability should not be viewed by others (who are also vulnerable, but often afraid of admitting it!) as giving them the right to injure us, to dispose of us, to abuse our bodies in situations of violence or of a Darwinist medical negligence, for instance.

In fact, Butler believes that *vulnerability* as such is prior to *experiences* of vulnerability, "even prior to individuation itself" (Butler 2004: 31), since we are bodies, exposed to "some set of primary others" (idem) from the beginning till the end of our lives. Quoting Butler: "we are vulnerable to those we are too young to know and to judge and, hence, vulnerable to violence; but also vulnerable to another range of touch, a range that includes the eradication of our being at the one end, and the physical support for our lives at the other" (idem).

Actually, Butler's starting question ("who is grievable?"), not only makes sense as a question about the "invisibles" that were victims of American bombings (as a consequence of 9/11), for instance. All her books about ethics and about its relation with bodies pose the same question about bodies that are considered as if they do not exist, that is, bodies that are rendered invisible: bodies from excluded people, from sexual, religious, and cultural minorities, bodies of old people, bodies that do not fit into the notion of "a human body", bodies which are hiding or hidden, giving, often secretly, a "shape to the water".[5]

However, invoking those bodies is not enough to include them into what Butler considers "an established ontology" (Butler 2004: 33). She calls for "an insurrection at the level of ontology" (idem), an ontology that starts by questioning what is real, whose lives are real, who has the right to be buried, who takes from others the right to be buried. "Invisible bodies", bodies that were turned into "invisible bodies", are already unburied, even if they are alive. And they may be considered "unburiable"[6] both in life and in death (Butler 2004: 34)—there is no room, no grave for them. And yet it is the recognition of "a common vulnerability", prior to the formation of the "I" that transforms vulnerability into this insurrect ontology, a "critical opening up of the questions" (Butler 2004: 33) that unburies[7] the invisible bodies and struggles to acknowledge in them the ontological right to be visible and to be buried, that is, the right to be humanized.

---

[5] For a possible interpretation of this "invisibility" see Guillermo del Toro's movie "The Shape of Water".

[6] Term used by Butler.

[7] Term used by Butler.

However, if we were able to grieve all human lives, and, on the other hand, if we were to become mired in "the sense of loss", that is, a passive and powerless sense, we would not be able to return to a sense of "human vulnerability" that would compel us to assume "our collective responsibility for the physical lives of one another" Butler 2004: 30). An active grieving turns grief into "a resource for politics"—a slow process by which we develop a point of identification with suffering itself, "an apprehension of a common human vulnerability" (idem). This means that in talking about this common human vulnerability there is still the need to take into account the existence of a "struggle for recognition".

The invisibilization of certain bodies is an expression of a "differential field of power", of "a differential operation of norms of recognition" (Butler 2004: 44). According to Butler, this does not negate the need to keep speculating about "the formation of the subject" (idem), since this would appear to be crucial in order to achieve a "theory of collective responsibility" (Butler 2004: 44) that will not mean a "radically inadequate care" (idem). One must acknowledge the existence of theories of power and recognition that foster dependence on a set of norms of recognition that may be impoverishing, violent or abusive, that may even put us in the position of being "the model of the human" (Butler 2004: 46). The risk in falling prey to this temptation is to adopt, once again, the position of those who decide on who is grievable and who is not.

In fact, Judith Butler's ontology does not only presuppose that there is an "I" who addresses others and/or that addressing others is always a successful endeavour. She considers that "what binds us morally has to do with how we are addressed by others in ways that we cannot avert or avoid; this impingement by the other's address constitutes us first and foremost against our will or, perhaps put more appropriately, prior to the formation of our will" (Butler 2004: 130). Butler distances herself from an understanding of "moral authority" according to a unilateral, more precisely, a modern notion of "autonomy" (even if she does not mention it explicitly in this piece of her text about "precarious life"). Quoting her again: "If we think that moral authority is about finding one's will, it may be that we miss the very mode by which moral demands are relayed. That is, we miss the situation of being addressed, the demand that comes from elsewhere, sometimes a nameless elsewhere, by which our obligations are articulated and pressed upon us" (Butler 2004: 130).

5 Lévinas and the Relevance of the Unavoidable Face

It is at this point that Butler introduces her own reading of Lévinas, more specifically, of his notion of "the face": an irreducible moral claim that comes from an external dimension, that we "do not ask for", "that we are not free to refuse" (Butler 2004: 131) unless we want to put ourselves in the position of a murderer or of an accomplice to murder.

Erasing the face of the other would mean to deny what Lévinas considered the "most basic mode of responsibility" (cit. in Butler 2004: 133). Commenting on

Lévinas' understanding of "the face", Butler writes: "the face exposes simultaneously what is precarious in another life or, rather, the precariousness of life itself" (Butler 2004: 134). However, this exposing of the precariousness of life, expressed in the irreducibility of a face, does not mean that this face would work like a mirror to us, in which we would "see" our precariousness, and then make an internal movement towards understanding the precariousness of the other. We are not at the centre of this reasoning: "the other" is the one who is at the centre, challenging us.

The radical challenge is to decide whether the precariousness of the other will lead me to fall into the temptation of killing him or will move me to make peace with him. At the origin of that decision, according to Lévinas and to Butler's reading of his philosophy of the face, is a voice that is not human. It is not the voice of a face that may be speechless, although bespeaking an "agony, an injurability".[8] The face of the other produces a struggle inside of us: a struggle between an impulse to kill, to take advantage of someone else's precariousness, in order to protect our own precariousness, and a prohibition against doing so. This voice does not belong to "the face". The voice that "bespeaks an agony" is the voice that "bespeaks a divine prohibition against killing" (Lévinas, cit in Butler 2004: 135). The face of the other "interrupts" the "narcissistic circuit" (Butler 2004: 138). It opens the possibility of an ethical thinking that recognizes vulnerability as an ontological condition that no one will be able to erase from the face of the Earth.

Butler's interpretation of the emphasis that Lévinas puts on "the face" and on the prohibition against killing is aware of the difficult questions it raises. One of them results from its frame: he speaks from the point of view of a dyad. However, in the sphere of politics, there are always more than two subjects. What if violence is structurally perpetrated against someone we love? How can we reconcile an ethical purpose of not killing with menaces against that someone? "Can I invoke the imperative to preserve the life *of the Other* even if I cannot invoke this right of self-preservation for myself?" (Butler 2004: 140).

It seems difficult to find all the answers formulated by Butler about this notion of "the face" and of the "Other" as an ethical *locus*. However, she thinks that Lévinas opens the way to think about "the relationship between representation and humanization" (PL, p. 140). It may well be "in the domain of representation where humanization and dehumanization occur ceaselessly" (Butler 2004: 140). With this sentence Butler opens an interesting discussion about the use of the face in a manipulative way within the media.

The main question is: "how do we come to know the difference between inhumane but humanizing face, for Lévinas, and the dehumanization that can also take place through the face?" (Butler 2004: 141). If we take the notion of humanization and dehumanization from the point of view of the possibility to gain representation, we can say that those who have the chance to represent themselves may have the temptation to decide on who will be treated as less human, or even who will be presented as dehumanized—in a nutshell: who will "disappear from our sight", who

[8] Term used by Butler.

will be made invisible. On the other hand, showing a face may also enact violence. Butler gives the example of the exhibiting of faces of the enemy as if they were "spoils of war". However, that is not Lévinas' point, since, for him, the human is not *represented by* the face. In Butler's words: "the human is indirectly affirmed in that very disjunction that makes representation impossible, and this disjunction is conveyed in the impossible representation. For representation to convey the human, then, representation must not only fail, but it must *show* its failure. There is something unrepresentable that we nevertheless seek to represent, and that paradox must be retained in the representation we give" (Butler 2004: 144). The human is beyond the representation we have and make of him: no picture, no photo is able to frame, to freeze a human being. No life, no grief, no pain can be captured through a picture. A picture is just a glance. Actually, sometimes, a picture is like a shot (in the sense of a photographic frame and in the sense of a shoot), that is, the moment that the camera starts rolling until the moment it stops, the continuous footage or sequence between two edits or cuts. A consummation of "our own inhumanity" (Butler 2004: 151).

6 Normative Commitments of Equality

In *Frames of War* (2009), Butler returns to the insufficiency of a "visual apprehension of a life" to form "a precondition for an understanding of the precariousness of life" (Butler 2009: 51). The book revisits the topics of "grievability"[9]and of "precariousness" already explored in *Precarious Life*, specifying some of them. The notion of precariousness that appeared in *Precarious Life* could be understood almost as a synonym of vulnerability. In *Frames of War*, Butler does not refer as often to vulnerability as she previously did and considers "precariousness" and "precarity" as "intersecting concepts" (Butler 2009: 25). We are all precarious, but there are "political induced conditions" in which "certain populations suffer from failing social and economic networks of support and become differentially exposed to injury, violence, and death" (idem), as a result of "an unequal distribution of wealth", and of "differential ways of exposing certain populations, racially and nationally conceptualized, to greater violence" (Butler 2009: 28).

In that sense, "precarity" maximizes precariousness, a feature that is common to all human beings, but that is exacerbated by structures of power. As already mentioned, "precariousness" is common to all human beings, it is "shared", while 'precarity' is socially induced. Butler calls for the introduction of "strong normative commitments of equality" and invites "a more robust universalizing of rights that seeks to address basic human needs for food, shelter, and other conditions for persisting and flourishing" (Butler 2009: 29).

[9]Term coined by Butler.

She links this appeal to the need for overcoming the induced precarity. Butler considers that the "the postulation of a generalized precariousness calls into question the ontology of individualism" (Butler 2009: 33). It is not enough to recognize that life is precarious and, therefore, it must be preserved. The conditions that "render life sustainable" (idem) should also be improved. By saying that, she is making a criticism and, at the same time, a political statement: an approach to precariousness that only understands this situation and this notion as an individual situation or condition may risk undermining the definition of political measures to ameliorate those situations where precariousness actually becomes "precarity". Precariousness cannot be used as a kind of "alibi" for social inequalities. Therefore, Butler's "ontology of the body" serves as a point of departure for thinking about social and political responsibility.

However, thinking of an ontology based on the body does not mean that it "persists" *per se*, since, in order to 'be", the body needs to "rely on what is outside itself" (Butler 2009: 33). The body is vulnerable, at the mercy of how sustainable or unsustainable the world is. The task of "responsiveness", of "responsibility", "is located in the affective responses to a sustaining and impinging world" (Butler 2009: 34). In short: "moral theory has to become social critique" (Butler 2009: 35). That is why Butler clearly calls for a notion of "responsibility" that is not only individual. We are bound to others: it is not possible to think of an "I" without "that world of others" (idem). At the end of the day: responsibility turns an "I" into a "we". Better still, it uncovers what we could call a relational ontology: if I exist, it is because there are others.

It is obvious that their existence may include the possibility of violence, of destruction, but it also includes the notion of a common "precariousness". Butler concludes: "In this sense, we are all precarious lives" (Butler 2009: 43). Nevertheless, there is still a question that actually is at the root of the ethics of care right from the beginning of its formulations[10]: how shall we consider each and every human being as someone we are responsible for? Butler goes even further and brings once more into her discussion "the regulatory power" that creates a divide between "those we are concerned with" and "those who are not even considered as human beings", those who are out of our sight, those "whose lives and deaths simply do not touch us, or do not appear as lives at all" (Butler 2009: 50)—a core topic of *Precarious Life*, as we have already seen.

How, then, is it possible to feel affected not only by the lives of those we love, but also by the lives of those we do not know? How are we to have an "affective response and moral evaluation" (Butler 2009: 50) that is aware of the existence of a previous framework which establishes who is "worthy of protection while others are not, precisely because they are not quite 'lives' according to prevailing norms of recognizability?" (idem). Butler considers that the only way to surmount these questions is "to perceive loss", to feel affected by loss. However, this will only be possible if "the social structures of perception", "the social supports for feeling"

[10] For a very interesting and enlightening history of the concept see Held (2006) and André (2021).

(Butler 2009: 53) recognize all human beings as precarious—all of them. And since we live and were born in a specific social frame, if in that frame all human lives are considered to be "grievable", it is possible to develop an "affective response to vulnerability, to the precariousness of precariousness—ours (and others' own bodies). In fact, in Butler's own words: "The boundary of who I am is the boundary of the body, but the boundary of the body never fully belongs to me. Survival depends less on the established boundary to the self than on the constitutive sociality of the body" (Butler 2009: 54).

But as much as the body, considered as social in both its surface and depth, is the condition of survival, "it is also that which under certain social conditions, imperils our lives and our survivability" (Butler 2009: 54). Nevertheless, as Butler states in *Notes Toward a Performative Theory of Assembly* (2015), living is always living in sociality. In that sense, "the life of the other, the life that is not our own, is also our life" (Butler and Athanasiou 2013: 108), from the start. We are "dependant on a world of others, constituted in and by a social world" (idem). This is why Butler thinks that "the ethical relation (…) is structured fundamentally by a mode of address: you call upon me, and I answer. But if I answer, it is only because I was already answerable, that is, this susceptibility and vulnerability constitutes me at the most fundamental level and is there, we might say, prior to any deliberate decision to answer the call. In other words, one has to be already capable of receiving the call before actually answering it. In this sense, ethical responsibility presupposes ethical responsiveness", that is, "ethics is always prior to the ontology of the ego" (Butler 2015: 110).

7 Final Considerations

Even though Butler has a positive understanding of vulnerability as something linked with the human condition as such, she also speaks of precarity as something that may be imposed on others through a process that renders them invisible by those holding the power to decide which vulnerability should be recognized and even given a "status". She recalls this negative side of precarity as an attribute imposed upon women that kept them from exposing themselves voluntarily (in the sense of participating) in public life (both from a political and a spatial point of view). The same happens (or happened) with colonized people who were considered lesser human beings, vulnerable in the sense of being dispossessed of human attributes. This negative or paternalistic understanding of precariousness may be misused in order to "guide" or oppress groups of people that may be considered incapable of ruling their own lives. And there is also a negative understanding of this condition that may, actually, be understood as "an advantage", for instance, in situations of war or of armed conflicts: vulnerability of the opposing side represents a chance to attack it.

Now, Butler considers that "vulnerability cannot be associated exclusively with injurability" (Butler 2015: 149), since it also means being /feeling affected by what happens in history: "Vulnerability may be a function of openness, that is, of being open to a world that is not fully known and predictable" (idem). From this point of view, everyone is vulnerable. However, that does not have to mean that we are caught in this vulnerability as in a trap. It is possible to mobilize people to recognize this human feature or to resist forms of vulnerability that are inflicted: "When the bodies of those deemed 'disposable' or 'ungrievable' assemble in public view (…) they are saying, 'we have not slipped quietly into the shadows of public life: we have not become the glaring absence that structures your public life'" (2015: 152). And this shows simultaneously precarity/vulnerability as something common and, on the other hand, as something inflicted, against which all of us (precarious, because we are human) can act, or, at least try to act.

In 2020, we were forcibly reminded that the possibility of being confronted with our precariousness in brutal ways was/is not something remote and unlikely. Rather, it means something that we cannot merely consider as an imponderable situation in our daily lives: the Covid 19 pandemics and the war in Ukraine caused by the Russian invasion—a conflict that started in February of 2022 and persists with no solution in sight—have taught us that. Not to mention climate changes and their impacts on vast regions of the world, with catastrophic consequences for the populations, especially for the most impoverished. In this sense, Kristen Dunphy's sentence ("There is a fine line between sanity and madness") quoted at the beginning of this text would seem to be translatable into the awareness of the "fine line" between "safety" and the "vulnerability" the world is experiencing.

Both situations demonstrate the volatility of life. Sometimes, vulnerability and precariousness reveal themselves in micro-experiences; other times, they are visible in macro-experiences, and on some occasions (nowadays, for instance!), in micro and in macro-experiences simultaneously. Nowadays, Butler's reflection on "whose lives can be marked as lives, and whose deaths will count as deaths" (Butler 2004: xxi) resonates in our minds with a different significance. She wrote these words with the American reaction to 9/11 in mind. Now, in the aftermath of Covid and in the middle of a war that has already had and will have serious impacts all over the world, we can keep asking ourselves whose lives can be marked as lives, and whose deaths will count as deaths.

Judith Butler's reflections and ethical questions seem to be very accurate and relevant to the development of an ethics of care in the present context of insecurity, lack of safety, traumatic and post-traumatic experiences. However, this is not just a "sad coincidence": Butler's thoughts about vulnerability consider that this represents our ontological condition. It is something that emerges in brutal and scary ways in particularly difficult times, but represents what we are: vulnerable.

Returning to the story of Wakefield mentioned at the beginning of this text: acknowledging our common vulnerability should lead us to try to find ways to dance on the edge of the abyss without slipping and falling into it while it is still possible to dance, since not falling does not mean, obviously, that we have the capacity to eradicate that possibility. In fact, the fall will be inevitable 1 day, otherwise we would be in a position of immortality.

Dancing on the edge of the abyss has much more to do, on one hand, with the awareness of what vulnerability means and, on the other hand, with seeing it as a sign of hope: Nik, the psychiatric nurse was not alone while dancing. He danced with someone else, someone who was also aware of the experience of being vulnerable and who nevertheless, made the effort of dancing with Nik. It was not a tragic dance, but a joyful one. At that precise moment, they danced while taking care of each other in order not to fall: care giver and care receiver were united in the recognition of their common vulnerability. Nik was not only a caregiver. He was also a care receiver. Are we not, all of us, in both positions?

References

André, João Maria. 2021. *Philosophy of care, advancing global bioethics* 16. Ed. J. Braga, M. Santiago de Carvalho, 19–37. https://doi.org/10.1007/978-3-030-75478-5_2.

Butler, Judith. 2004. *Precarious life: The powers of mourning and violence*. London/New York: Verso.

———. 2009. *Frames of war: When is life grievable?* London/New York: Verso.

———. 2015. *Notes toward a performative theory of assembly*. Cambridge: Harvard University Press.

Butler, Judith, and Athena Athanasiou. 2013. *Dispossession: The performative in the political. Conversations with Athena Athanasiou*. Cambridge: Polity Press.

Dunphy, Kristen. 2020. *Wakefield: There's a fine line between sanity and madness*. Jungle Entertainment/Australian Broadcasting Corporation. https://tapintomentalhealth.com/wp-content/uploads/2021/03/Wakefield-Press-Kit_18.02.21.pdf. Accessed 8 Sept 2021.

Engster, Daniel. 2019. Care ethics, dependency, and vulnerability. *Ethics and Social Welfare* 13 (2): 100–114. https://doi.org/10.1080/17496535.2018.1533029.

Held, Virginia. 2006. *The ethics of care. Personal, political, and global*. Oxford: Oxford University Press.

Miller, Sarah Clark. 2020. From vulnerability to precariousness: Examining the moral foundations of care ethics. *International Journal of Philosophical Studies* 28 (5): 644–661. https://doi.org/10.1080/09672559.2020.1804239.

Chapter 7
The Soft Call of Memory: On the Need to Care for Our Elders

Ana Falcato

> *Care, as a primordial structural totality, lies 'before' ['vor'] every factical 'attitude' and 'situation' of Dasein, and it does so existentially a priori; this means that it always lies in them. So, this phenomenon by no means expresses a priority of the practical attitude over the theoretical. When we ascertain something present-at-hand by merely beholding it, this activity has the character of care just as much as does a 'political action' or taking a rest or enjoying oneself. 'Theory' and 'practice' are possibilities of Being for an entity whose Being must be defined as 'care'.*
>
> (Martin Heidegger, *Being and Time*, §41, 194: 238)

Abstract In section 41 of *Being and Time*, Martin Heidegger specifies what he takes to be the fundamental mode of being that Dasein, or Being-There, has as its own-most essence: care (*Sorge*). Heidegger defines 'care' in relation to anxiety and its capacity to disclose one's ultimate worldly trajectory as a meaningful, always already anticipated *whole* (cf. Heidegger, Being and time (trans: John Macquarrie and Edward Robinson). Blackwell, London, 1962, §41). By this he does not mean that care involves reaching out to others or fulfilling a social, humanist or cross-generational duty. In spite of his professed wish to keep the Analytic of Dasein separate from a philosophy of the subject, Heidegger's core notion of 'care' is self-referring and deeply egotistical. In this chapter, I defend an altogether different framework for thinking about care, grounded in a deep appeal to our earliest memories and the intimate world in which they are anchored—a world which, under the care of our elders (our grandparents and other members of their generation), helped to form us as the types of beings we now are. The formative bond we share with those who once cared for us, who have in the meantime become fragile and dependent, imposes a duty of care that is not rooted in a concern for *my* future or antici-

A. Falcato (✉)
Universidade Nova de Lisboa, Lisbon, Portugal

J. Braga (ed.), *Ethics of Care*, Advancing Global Bioethics 20,
https://doi.org/10.1007/978-3-031-73608-7_7

pated lifespan. This chapter thus seeks to refute Heidegger's treatment of care, which is one of the core notions of his thought.

Keywords Care · Childhood memory · Duty of care · Nostalgia

1 Introduction: Getting the Time Right

When the COVID-19 pandemic broke out at the beginning of 2020, national governments, national and international health organizations, and local authorities launched a massive information campaign about the virus, its transmissibility, daily contagion figures, death tolls, measures to minimize risk, etc.—in short: a wealth of chaotic orders and recommendations—in which one phrase was paramount: the '*new normal*'.

The 'new normal', several spokespeople claimed, encompassed a set of social habits and behaviours, which were likely there to stay, and which were designed to minimize infection rates and ease the burden on national health services. Among the many measures adopted in most countries was a civic (and widely policed) obligation to stay at home and work remotely. Companies and institutions were forced to adopt a set of social and economic measures meant to 'protect' work and health and to reduce the impact of the virus on societies and businesses.

Both the duty to stay at home and remote work prevailed for many months in most countries and were reinstated several times after brief suspensions. For most people, such state-imposed measures caused unbearable levels of stress and psychological strain, but they also had practical advantages, which tend to go unspoken. The duty to stay at home meant we could care for the people we most cherish in life, including the elderly, who were much more dependent on relatives than on state or privately funded institutions, some of which have a record of abusing those who can no longer defend themselves.

In this chapter, I argue that one practical side effect of the globally enforced lockdowns during the COVID-19 pandemic was the sudden availability of younger, healthier people to care for the elderly—who, in our globalized capitalist society, are generally unprotected by the state and often abandoned by younger relatives, who must tend to their 'normal' daily work and routines. This, I claim, was an utterly *welcome* new normal.

In order to back up this general claim, I will examine one of the most venerable philosophical defences of care as an ontological megastructure that defines the human being: the argument provided by Martin Heidegger (1962 [2013], §41ff). The aim of this examination is to identify where this model founders, both conceptually and in practical terms. Thus, in the following pages, I lay out in detail the overall features of Martin Heidegger's defence of *care* (*Sorge*) as the fundamental ontological disposition of Dasein, or Being-There (the kind of entity we ourselves are). Special attention will be given both to the critical point within *Being and Time* at which this core ontological notion first appears and to providing an outline of

Heidegger's specific understanding of care. In unfolding the conceptual nuances that mark Heidegger's concrete definition and step-by-step explanation of *Sorge*, I will close by concluding that his definition is biased and unsatisfying when read against a broader understanding of care that covers all kinds of human experience, including the need to reach out to other human beings, toward whom one may have a real duty of care.

2 Martin Heidegger on *Sorge* and *Fürsorge*

The most consistent and ontologically justified modern philosophical defence of care can be found in the work of Martin Heidegger, for whom care (*Sorge*) is the fundamental mode of being of Dasein (Being-There). Heidegger specifies what he means by 'care' in relation to anxiety and its capacity to disclose one's ultimate worldly trajectory as a meaningful, always already anticipated whole, ahead of which we place ourselves (cf. Heidegger 1962, §41). With his well-known taste for the formulaic, Heidegger defines *care* as the fundamental behaviour of Dasein as 'Being-ahead-of-oneself' (Heidegger 1962 [2013], §41) and reclaims *anxiety* as the key emotional disposition in which such an ontological mega-structure imposes itself upon one's experience of oneself.

In paragraph 41 of *Being and Time*, we find the following explanation:

> Since our aim is to grasp the totality of this structural whole ontologically [i.e, of Dasein or Being-There], we must first ask whether the phenomenon of anxiety and that which is disclosed in it, can give us the whole of Dasein in a way which is phenomenally equiprimordial, and whether they can do so in such a manner that if we look searchingly at this totality, our view of it will be filled in by what has thus been given us. The entire stock of what lies therein may be counted up formally and recorded: anxiousness as a state-of-mind is a way of Being-in-the-world; that in the face of which we have anxiety is thrown Being-in-the-world; that which we have anxiety about is our potentiality-for-Being-in-the-world. [...] The formally existential totality of Dasein's ontological structural whole must therefore be grasped in the following structure: *the being of Dasein means ahead-of-itself-Being-already-in-(the-world). This Being fills in the signification of the term 'care' [Sorge], which is used in a purely ontologico-existential manner.*[1] From this signification every tendency of being which one might have in mind ontically, such as worry [*Besorgnis*] or carefreeness [*Sorglosigkeit*], is ruled out (Heidegger 1962 [2013], §41, 191–193).

Unlike belief, care for Heidegger is not a mode of behaving or reaching out to others or—even less so—of specifically fulfilling a social, humanist[2] or cross-generational

[1] My italics.

[2] As is well known, Heidegger despises the notion of 'humanism', which he claims to be tied to a traditional view of metaphysics (see Heidegger 1949 [1977]). In the *Letter on Humanism* (Heidegger 1977), Heidegger criticizes the French existentialists' views on humanism, claiming that the notions of 'humanism' and 'humanitas' are unfit to account for what is specific to the form of being of human beings (in *Being and Time*, Heidegger had already rejected the very notion of the 'human being' and reclaimed the formulaic term 'Dasein' to replace it). His rejection of humanism as an all-encompassing notion is based on his own understanding of both traditional

duty. For the supposed mode of caring *for others*, which in fact encompasses Being-With other human beings, Heidegger reserves the term *Fürsorge*, which is usually rendered in English as 'solicitude'[3]—although it literally means to 'care for'.[4]

In spite of his professed wish to keep the Analytic of Dasein separate from a contemporary philosophy of the subject, Heidegger's core notion of 'care' is deeply *egotistical*; as I will show in what follows, in its very definition in *Being and Time* the notion is unequivocally self-regarding. Although Heidegger would likely dismissively label such critical terminology, including the very notion of the 'ego' (and its use as part of a rebuttal of his ontological model), as 'psychological' or 'anthropological' at its roots, the meaning Heidegger conveys throughout his description of care as the fundamental mode of being of Dasein is rendered clear as he proceeds with his analysis, repeatedly returning to the core notion used to characterize the kind of Being that we ourselves are—namely, *Sorge* or *care* (cf. Heidegger 1962 [2013], §§ 41, 42, 64).

My main concern in this chapter is to defend an altogether different experience and *praxis of care*, grounded in a deep appeal to our earliest memories and to the intimate world in which they are anchored—a world which, under the care of our elders (our grandparents and other members of their generation), helped to form the human being we once were and still are. The formative bond we share with those who once cared for us, who have in the meantime become fragile and dependent, imposes a *duty of care* that is anything but burdensome or rooted in a concern for *my* future or anticipated lifespan. In fact, the need to care for our elders *now* is yet another common enterprise—just as our first shared experiences were joint ventures—and the daily routine of helping and assisting, grounded in a common emotional background, is one of joy and shared purpose, our different stations and critical needs in life notwithstanding.

This fundamental experience, enjoyed by emotionally responsible caregivers, will be described in further detail below, although it is utterly absent from

metaphysics and the core temporal structure he ascribes to the being of Dasein: *ek-sistence*. Sartre, however, who is the main target of Heidegger's criticism in the *Letter*, used different terminology in his ontology of the human subject (see Sartre 1943 [1992]) and spoke of the three *ex-stases* that define the being of the *Pour-Soi* (cf. Sartre 1992: Part Two and Part Three).

[3]Cf. Heidegger 1962 [2013], 121: 'If Dasein-with remains existentially constitutive for Being-in-the-World, then, like our circumspective dealings with the ready-to-hand within-the-world (which, by way of anticipation, we have called "concern"), it must be interpreted in terms of the phenomenon of *care*; for as "care" the Being of Dasein in general is to be defined. Concern is a character-of-Being which Being-with cannot have as its own, even though Being-with, like concern, is a *Being towards* entities encountered within-the-world. But those entities towards which Dasein as Being-with comports itself do not have the kind of Being which belongs to equipment ready-to-hand; they are themselves Dasein. These entities are not objects of concern, but rather of *solicitude*.'

[4]The source of the difficulties associated with interpreting 'Fürsorge' may lie in its most current English translation—which most English-speaking Heidegger scholars have found very difficult to avoid (see John Macquarrie and Edward Robinson, Heidegger 1962 [2013], p. 157). Whereas a literal interpretation of *Fürsorge* seems more adequate to the specific purposes of the interpretation proposed here, it is undeniable that Heidegger stakes his claims in the all-encompassing notion of care (*Sorge*), taking *Besorgen* (concern) and *Fürsorge* (solicitude) as derivatives.

Heidegger's many descriptions of the core concept of *Sorge* in *Being and Time.* The central claim at the heart of this chapter is that, given the way in which post-industrial European societies have evolved in relation to ageing and life-expectancy (in a context with a substantially ageing population and ever-scarcer social resources), it is essential for those seeking to provide a *useful philosophical account* of care to first (i) *unfold theoretically* and then ultimately (ii) *discard* the Heideggerian concept of care once its fundamental flaws (even in its own terms) have been revealed.[5] It would be pointless, of course, to reject a strictly ontologico-theoretical notion (of care) for the mere sake of replacing it with a socially responsible one; in doing so we would be trading at different levels and making progress in none. My point is rather that, when properly understood, *Being and Time*'s own definition of care as a fundamental existential framework leads to deeply misguided practical and even *ontological* conclusions.

To better explain and support my main critical point, let us consider the first description of the notion at the heart of this chapter—*Sorge* or care—in *Being and Time*. It first appears in paragraph 41 (quoted above):

> The formally existential totality of *Dasein's ontological structural whole* must therefore be grasped in the following structure: the Being of Dasein means ahead-of-itself-Being-already-in-the-world as Being-alongside-(entities encountered within-the-world). This Being fills in the signification of the term '*care*' [*Sorge*], which is used in a purely ontologico-existential manner. From this signification every tendency of Being which one might have in mind ontically, such as worry [*Besorgnis*] or carefreeness [*Sorglosigkeit*], is ruled out. *Because being-in-the-world is essentially care, Being-alongside the ready-to-hand could be taken in our previous analyses as* ***concern,*** *and Being with the Dasein-with of Others as we encounter it within-the-world could be taken as* ***solicitude*** *[Fürsorge]* (Heidegger 1962 [2013]: § 41, 237).[6]

In purely ontological terms, this means the following: Because the kind of being we ourselves are has as its fundamental mode of being *being-ahead-of-itself* and has *being with things* and *being-with-others* as co-constitutive possibilities, it has care as a **structural nest of attitudes**, which encompasses the three modes of *inter-* and *self*-referential relationality. Thus, care is the core notion and core behaviour of Dasein, which can be divided into *concern* (care for things encountered at hand) and *solicitude* (care for other entities with the same ontological structure as our own, i.e., other Daseins, or in proper German, *Daseinende*).

At its roots, according to Heidegger, care comes from the fundamentally temporal mode of being-ahead-of-oneself and caring and looking for(ward) to what is outstanding. Its manifestation in the context of other human beings is *solicitude.* Thus, the latter is also a derivative experience (mapped onto a subsidiary notion), as it emanates from the structural and primary form of care as being-ahead-of-oneself (in each case, *my* being ahead of myself).

---

[5] I.e., the terms in which the notion of *Sorge* is framed within the conceptual narrative of *Being and Time*, and what lies behind a proper grasp of what Heidegger has in mind with this mega *existentiale*, as developed in his *magnum opus*.

[6] My italics.

As I wish to argue, however, the idea of 'solicitude' is a misnomer for something to which the term also refers within the categorical map of *Being and Time*—namely the *care* of (also) fragile and dependent human beings. This concept need not be derivative and fundamentally anchored in the first and all-encompassing notion (and experience) of care, conceived of by Heidegger as the outcome of an experience of anxiety and of how it discloses Dasein to itself as a meaningful whole (cf. Heidegger 1962 [2013]: §40).[7]

Care for the elderly is, of course, a paid profession for some and a vocation for many. This is not the point I am stressing here. What I have in mind is rather the direct counterpart of the idea of care as the enactment of one's core existential possibility, a counterpart that does not derive from *being ahead of oneself and still less subsumes care for others as a mode of an essential self-structure.* It is also important to stress that I am not alluding to something akin to the revelation of a religious or humanistic vocation. What, then, am I referring to? The following first-personal example will help to clarify the core affective bond of which I am speaking, as well as the notion of a *duty* of care.

The beginning of one's life may well be a mystery to one, and one's earliest recollections can be distorted in many ways. Nevertheless, there are factical human presences that one *never* forgets and whose core impressions upon one's emotional and sensitive apparatus are too clear and long-lasting to be fabricated. Growing up in the last decades of the twentieth century, in a poor but peaceful country—Portugal in the last years of the Cold War—under a routine of daily work constraints that absorbed parents' time, children were often left in the care of the older generation (grandparents, and even great-grandparents). Strong emotional bonds inevitably developed, giving the child a sense of meaning, purpose and belonging to a shared world. More than 30 years have passed since then, and these early caregivers have grown old, frail and physically dependent. The constant rush—the economic and political cataclysm—that marks the present epoch has left young adults little time to look after those who first looked after them. What has been lost as a result is a golden opportunity to celebrate and meaningfully recapture a time when things were easier for both generations.

Caring for the elderly *now*—rather than leaving them at the mercy of predatory and abusive institutions—is more than an intergenerational (often burdensome)

[7] Cf. Heidegger 1962 [2013], 184–188: 'One of Dasein's possibilities of Being is to give us ontical "information" about Dasein itself as an entity. Such information is possible only in that disclosedness which belongs to Dasein and which is grounded in state-of-mind and understanding. How far is anxiety a state-of-mind which is distinctive? How is it that in anxiety Dasein gets brought before itself through its own Being, so that we can define phenomenologically the character of the entity disclosed in anxiety, and define it as such in its Being, or make adequate preparations for doing so? [...] That in the face of which anxiety is anxious is nothing ready-to-hand within-the-world. But this "nothing ready-to-hand", which only our everyday circumspective discourse understands, is not totally nothing. The "nothing" of readiness-to-hand is grounded in the most primordial "something"—in the *world*. Ontologically, however, the world belongs essentially to Dasein's Being as Being-in-the-world. So if the "nothing"—that is, the world as such—exhibits itself as that in the face of which one has anxiety, this means that *Being-in-the-world itself is that in the face of which anxiety is anxious.*'

duty. It is an *opportunity*, indeed the last available opportunity, to make sense of a shared story, to put one's youth, one's physical and moral vitality, to good use and to learn more about oneself and all preceding generations, whose labour, privations and personal and collective stories have brought us to where we (previous and current caregivers) now are.

This type and practice of caring has nothing whatever to do with Heidegger's seemingly ontologically irreplaceable and all-encompassing notion of *Sorge* as the core form of Dasein's behaviour toward itself once it realizes its fundamental way of being as *being-ahead-of-itself.* It is least of all anything resembling *solicitude* as a co-constitutive mode of the being of Dasein as (always) Being-with-others, not only because such a term may have *aesthetic* more than ontological (let alone ethical) connotations (which is what is needed here, that is, *ethical connotations and a thoroughgoing ethical shift in one's life*) but because the way in which the term is framed by Heidegger in *Being and Time* entails that, deep down, what the notion of care properly amounts to is a *caring for oneself.*

My example, *qua* typical life experience in the time of a global pandemic, like so many others we can think of nowadays (the war that has broken out less than 5000 km away from the place where I am writing, bringing with it the cataclysmic threat of nuclear disaster, the unprecedented refugee crisis that continues to be poorly dealt with by European regimes, etc., etc.) disproves the universal validity of the structural notion of *Sorge* as presented in *Being and Time*. *Sorge* (*care*), as a core existential disposition of Dasein, cannot function as a universal disposition or fundamental mode of being of the entity we ourselves are when its step-by-step descriptive content is so staunchly disproved by a single critical instance of an (ontical) experience of care.

Put into context, Heidegger's definition of care not only loses the ontological force it is meant to have but is shown to be biased toward a care of the self, on the one hand, and hostage to the stage in the narrative of Dasein in which it appears in *Being and Time* (§§ 40, 41), on the other. Being strictly *me-regarding*, it is dangerous in its straightforwardly practical implications, to the extent that it precisely discourages other forms of care, which are much more intuitive and socially urgent. Despite feasible interpretations of the German term that Heidegger uses to characterize the split notion of *Fürsorge* (as a 'branch' of care, applied to our dealings with other human beings), which tend to be associated with welfare work or 'domestic arrangements for care',[8] such commonsensical and literal interpretations of the orig-

---

[8]The Macquarrie and Robinson translation of *Being and Time* contains the following note on the difficulties of rendering *Fürsorge* in English: 'There is no good English equivalent for "Fürsorge", which we shall usually translate by "solicitude". The more literal "caring-for" has the connotation of "being-fond-of", which we do not want here; "personal care" suggests "personal hygiene"; "personal concern" suggests one's personal business or affairs. "Fürsorge" is rather the kind of care that we find in "prenatal care" or "taking care of the children", or even the kind of care that is administered by welfare agencies. Indeed, the word "Fürsorge" is regularly used in contexts where we would speak of "welfare work" or "social welfare". This is the usage which Heidegger has in mind in his discussion of "Fürsorge" as a "factical social arrangement"' (cf. Heidegger 1962 [2013]. 121: 157).

inal term for 'solicitude' cannot satisfy us in light of what we have seen above. For the problem is not mainly with the unfit aspects of the chosen English rendering of the German '*Fürsorge*' but with the framing of *Sorge* itself in *Being and Time* and how the experience related above, *qua* a primary instance of care, utterly fails to match it. Thus, what we have been presenting is a fundamental experience *of care* (not solicitude, nor even '*Fürsorge*' in the sense of social work) that in no way corresponds to what Heidegger means by *Sorge*[9] in his primary conceptual account. In a non-trivial sense, the problem we have identified has the following basic structure: we need a notion of care that has sufficient semantic and philosophical nuance to express what is going on in this fundamental human experience, but it cannot be Martin Heidegger's, for the reasons pointed out above.

3 Concluding Remarks

I would like to end this brief chapter by returning to the idea that first gave rise to these reflections: the plural notion of '*roads* to care', its multifarious resonances, and critical, philosophically informed approaches to what care may imply and demand of us—subjects, citizens and moral agents living in the twenty-first century, who have recently undergone the radical existential strain of surviving a global pandemic. I also wish to point out the great disparity, at the level of analytic detail and interpretive framework, between the classic ontological conception of care in relation to human temporality (which is what is at stake in *Being and Time*, with a special emphasis on paragraph 41) and the political, social and economic decisions (as enforced during the global COVID-19 pandemic) that created a context in which care—more concretely, healthcare—was called for and institutionally enforced.

At the time of writing *Being and Time*, Heidegger was not thinking of something akin to a health emergency (perhaps not even a war), and we are in no way compelled to read the words of a twentieth-century philosopher in order to understand our present context and what it may require in terms of care or care measures. In this sense, reading and re-interpreting what Martin Heidegger had to say about care is no more than an academic exercise. This exercise bears conceptual fruit, however, which may be put to good use in both understanding and re-evaluating what Heidegger wrote about one of the key concepts of his first philosophical project and in determining whether it makes sense to apply such reflections to a specific social and political context that demands more thought and consideration.

In addition, and in relation to everything that has been written above, I would like to make a minor adjustment to the social and political strand of the argument—i.e., regarding what care may come to mean and imply in times of a health emergency.

[9] Another, related aspect is that *Sorge*, in common German usage, does not generally mean what Heidegger intends it to mean in his jargon. It has become trivial for German native speakers to discard and mock supposedly primordial aspects of the use of German terms in the Heideggerian lexicon as distortions of proper usage.

This is the observation that *the deep vulnerability of human beings living under conditions of impoverishment and existential uncertainty is far from being a byproduct or social and societal outcome of the COVID-19 pandemic*. For many years now (it is certainly hard to venture exact figures or a precise date, although philosophers, historians and writers alike have often identified the last years of the Cold War as a turning point), human relationships—social, institutional, labour and economic structures and forces—have been undergoing deep, often imperceptible changes, which are anything but normal by any reasonable standards. The so-called 'new normal' heatedly discussed in the media during the COVID-19 pandemic is therefore all but new.

The intergenerational features highlighted in this chapter, however—especially given their urgency, even at the political level—have rarely been seen in common democratic life in Western countries and societies. On the one hand is the pressing social and infrastructural preoccupation with the older generation that has characterized most European governments in the last few decades, to the point of holding research funding in all disciplines hostage to socially responsible strategies[10] (some of them transgenerational). Another, completely different development is the state-imposed prohibitions that were brought about by national and non-national lockdowns, which brought people face-to-face with new routines and possibilities—as well as unprecedented levels of stress and mental strain. The sort of spatial and personal flexibility that was still allowed in contexts of funded academic research gave some (the number of people engaged in such research has been diminishing every year, as the research community is well aware) the opportunity to finally do what they've always wanted to do: *to take care of those who first took care of them*—not out of anything resembling solicitude (*Fürsorge*), but simply out of love.

References

Heidegger, Martin. 1962 [2013]. *Being and time*. Trans. John Macquarrie and Edward Robinson. London: Blackwell.

———. 1977. *Basic writings*. Ed. David Farrell Krell. New York: Harper & Row.

Sartre, Jean-Paul. 1943 [2014]. *L'Être et le Néant: Essai d'ontologie phénoménologique*. Paris: Gallimard.

———. 1992. *Being and nothingness: A phenomenological essay on ontology*. New York: Washington Square Press.

[10] Cf. the Sustainable Development Goals (SDGs), also known as the Global Goals, adopted by the United Nations in 2015 as a universal call to end poverty, protect the planet, and ensure that all people will enjoy peace and prosperity by 2030 (https://www.undp.org/sustainable-development-goals?utm_source=EN&utm_medium=GSR&utm_content=US_UNDP_PaidSearch_Brand_English&utm_campaign=CENTRAL&c_src=CENTRAL&c_src2=GSR&gclid=EAIaIQobChMIscyFg7bW_AIVFPd3Ch0HLwV1EAAYAiAAEgK2XfD_BwE). In Portugal, for instance, publicly funded research has been made conditional on an explicit commitment to at least some of the goals of the 2030 UN Agenda.

Part II
Care in Practice

Chapter 8
Care and the Market: Can We Consume with a Concern for Others?

Estelle Ferrarese

Abstract One question regularly recurs concerning an ethics of care, which is how it can be applied beyond the intersubjective relationship, or otherwise than within the *caregiver/care receiver* dyad. To the extent that it presumes affects, or an attentiveness to the particular or the individual, is it possible to show concern collectively and bring it to bear on an unknown other? This inquiry is generally approached through the prism of the institutionalization of an ethics of care, of its compatibility with the scale of the state and its logic of universalization. But the question I wish to address here is that of the market. Can care, which I will call concern for others, be deployed on and with the market? This is the proposition advanced by contemporary consumer ethics and the one I would like to critique in this paper. After drawing up a cartography of the objects with which contemporary ethical consumption is concerned, I will show that there is a vainness to ethical consumption as a critique of capitalism insofar as it centers on immoderation. I will then highlight what the market does to care by translating it into the language of prices.

Keywords Care ethics · Market · Ethical consumption · Capitalism · Fair price

One question regularly recurs concerning an ethics of care, which is how it can be applied beyond the intersubjective relationship, or otherwise than within the *caregiver/care receiver* dyad. To the extent that it presumes affects, or an attentiveness to the particular or the individual, is it possible to show concern collectively and bring it to bear on an unknown other? This inquiry is generally approached through the prism of the institutionalization of an ethics of care, of its compatibility with the scale of the state and its logic of universalization. But the question I wish to deal with here is that of the market. Can care, which I will call concern for others, be

E. Ferrarese (✉)
Université de Picardie Jules Verne, Amiens, France
e-mail: estelle.ferrarese@u-picardie.fr

J. Braga (ed.), *Ethics of Care*, Advancing Global Bioethics 20,
https://doi.org/10.1007/978-3-031-73608-7_8

deployed on and with the market? This is the proposition advanced by contemporary consumer ethics and the one I would like to critique in what follows.[1]

A cosmetics brand claims, on the packaging of its products, that they are "100% paraben and cruelty free." Marketing textbooks ponder over how consumers calculate utility and digest information leading them to opt for products made in equitable conditions. A wine specialist sets out at length on its website how the vast majority of its deliveries arrive by bicycle or public transport. A pandemic persuades the inhabitants of large cities to buy "local" products from "small" businesses.

These statements express so many new ways to relate to the market and to commodities. The practices themselves have even come to constitute self-evident facts in political discourses and the calls made in social networks. They allow for radical shifts in position with respect to capitalism. The point is indeed sometimes to amend, sometimes to avoid, and sometimes to crack its logic of accumulation. It is to amend, insofar as one strives to correct the market mechanism for determining price to the extent that it works against the producer. It is to avoid when entire networks of production and consumption are organized outside of the rhythms, mediations, and categories of capitalist exchange. And it is to crack when it is hoped that reforming consumption will lead to a reform of production and the wage system. We have only to note that in no case is the goal to place certain things (such as the earth) outside the market. Nor is it to conceive and establish a form of collective property. Nor, lastly, is it to abolish the market as such. On the contrary.

In their diversity, all these practices, and the structures in which they sediment, aim to inject norms, rules and moral gestures into market exchanges, which contemporary capitalism is charged with having hollowed out. In other terms, what defines this mode of ethical consumption is, more than a particular practice, a precise organizational form, or objective characteristics, the *intention* that informs it.

More exactly, the mark of ethical consumption lies in the *concern for others* shown or alleged to have been, in the act of purchase. Relating to the other through commodities does not exclude, but on the contrary demands, an attention to the needs and suffering of the other, coupled with a felt injunction to respond to it; in other terms, care is tasked with taming, or correcting, market relations. The enemy is not so much the market per se, its laws and its automatisms, but the indifference and coldness of its actors. The point is to show a concern about the effects and gestures performed in exchange, and to strive to master these effects, or in any case to steer them in a particular direction.

The other for which concern is shown here may be singular or collective, human or non-human, while the worry displayed in ethical consumption may bear on the vulnerability of life, or form of life, on the threat bearing down an animal, a species, or the environment in a broadly defined sense.

As a set of practices that manifests and actualizes an attentive concern for others and the wrongs that can be occasioned, ethical consumption sketches a moral life

[1] This article is an overview of the introduction and chapter 1 of my book *Le marché de la vertu. Critique de la consommation éthique* (Ferrarese 2023).

rooted in the every day, one that, far from finding its privileged expression in some dazzling act, resides in repeated, banal, and minute actions. It presents a conception of moral life that resembles an ethics of care. To start with, however, we ought to introduce a qualification: as these practices have an essentially negative moral texture, they are about abstaining from the inflicting of certain wrongs on others who are known to be vulnerable to one's behaviours. This can be clearly seen in instances when the point is to refuse to buy a commodity produced in questionable social or environmental conditions, to avoid consuming a product during specific periods, but is also true of practices where you agree to pay a "fair" price. For you thus abstain from profiting and taking advantage of your position in a system of exchange structured to enable you to do so. Otherwise said, if care ethics presents an ardent disposition, an active attitude—that of attention, of concern—as the wellspring of moral life, the gesture to which it enjoins you in ethical consumption is utterly classic and liberal. This ethical attitude does not demand a positive obligation to take responsibility for some vulnerability, to realize an act, or to engage actively in helping others; instead, it consists in abstaining from inflicting damage. The sort of wrong contemplated arises not from inaction but from (hostile or insensitive) actions, which you must only desist from doing.

The fact remains that ethical consumerism, or ethical consumption, seeks to express a concern for others. This concern is posited from the start as a posture whose moral value is tied to the *form of the world that would ensue*, and it is in this light that it ought to be examined. The critique that I seek to develop thus does not aim at the modesty of these practices in relation to market capitalism, or even market society. Tenants of an ethics of care actively assume this stance of moderation. Nor does it bear on the ambiguity of these practices. For we must leave aside the question of impurity regarding the concern for others as manifested in ethical consumption, that is, the suspicion that it is entangled with less noble motives. Empirical research has firmly established, for example, that "responsible" buying is also perceived as more likely to guarantee a product of superior quality to those acquired in conventional trade.[2] The critique of ethical consumption that I undertake aims at its effects on the social totality, and leaves the intensity and sincerity of the individuals involved in these practices in the background.

If we want to assess the general form of the world that results from a concern for others on the market, then the malformations, deformations or saliencies that result from this collective exercise of care are phenomena that ought not to be overlooked. Accordingly, for example, since the early 2000s, the demand for green energy has seen foreign companies acquire land on a massive scale in various regions of the world for the production of biofuel. This development has led to greater levels of

[2] According to numerous studies, the motivations of buyers of fair-trade products are of three types: first, consumers buy fair trade products in order to establish greater equity in economic relations; second, they do it in order to protect the environment and thus their health—a shift that is not insignificant from an ethical point of view; and, finally, they do so to obtain good quality products. Of these three, the argument about quality is the one most often put forward by buyers of fair-trade products seeking to justify their choice (Dubuisson-Quellier 2009: 49).

toxicity in the soil and water in zones that surround the plantations, and also sometimes works to "transform sovereign national territory into a far more elementary condition—land for usufruct" (Sassen 2014: 82). Concern for *an* other, then, often occasions a multitude of wrongs against other "others." Similarly, instead of replacing the fossil fuels that they compete against, renewable energies have simply added to them. Economists often speak of a rebound effect: because energy-saving light bulbs require less electric current than incandescent light bulbs, one installs more of them. The reduction in the carbon footprint brought about by the pandemic immobilization of the planet was cancelled as the increased use of digital technologies caused a gigantic rise in energy consumption. And it may well be that developing synthetic meat from cells and artificially produced proteins with the goal of avoiding animal cruelty weighs negatively on the carbon footprint of a meal. Given the form that they stamp on this world, these phenomena cannot be dismissed as simply having "fallen short," as blunders or lapses. Instead, they must be understood as symptoms of a "disadjustment," of something ill-founded in the "responsible" practices from which they stem. By underestimating the systemic functioning of the market economy, these practices were in fact destined to fail.

The view I adopt in this critique stems from the Frankfurt School critical theory analyses of exchange relations and descriptions of forms of capitalist life, in particular those by Theodor W. Adorno.

It is indeed in Adorno that we find a brief a priori challenge to the idea of frugality, which he makes in a famous aphorism in *Minima Moralia* that closes with the affirmation according to which there is no true life in a false world: "consumer goods have become potentially so abundant that no individual has the right to cling to the principle of their limitation." The idea of moderation no longer amounts to much when "private property no longer belongs to anyone" (Adorno 2006: 39); the form and volume of the capitalist appropriation of things is such that it cannot be corrected or transformed through individual action, or even with the addition of several others. Beyond the purpose of its endeavour, namely sobriety, ethical consumption is also targeted by Adorno for the *individual* manner that underlies it. Insofar as it suggests a diagnostic of the present day, my critique of ethical consumption, which borrows from Critical Theory in general, and Adorno in particular, can only proceed through additions and alterations. It nevertheless remains faithful to this tradition since it is motivated by the project to bring to light the social totality that ethical consumption actively reproduces (and does not simply leave intact).

Indeed, the point cannot simply be to show that ethical consumption, despite or thanks to its desire for market correction, actually sustains this same market. For it is clear that a commodity purchased in the framework of responsible buying is no less of a commodity, that ethical consumption *creates* new markets, and that fair trade sustains consumption. While this is no doubt true, my critique aims at a different level.

It bears on the effects of ethical consumption on an economy whose categories, grids of apprehension, movement, and temporality *exceed* the mechanisms of production and exchange, as well as the processes of valorization and exploitation. It aims to bring to light how ethical consumption participates in a social totality modelled on exchange. It is not so much the market's specific productions and contents that interest me here, as its *forms*, to the extent that they affect and pervade all spheres of life.

After drawing up a cartography of the objects with which contemporary ethical consumption is concerned, I will show that there is a vainness to ethical consumption as a critique of capitalism insofar as it centres on immoderation. I will then highlight what the market does to care by translating it into the language of prices.

1 The Concern for Others. Who Are These Others? What Is the Concern?

Practices of ethical consumption and their accompanying discourses establish a regime that distributes acts according to their legitimacy or illegitimacy, and ensure a—necessarily asymmetrical—division of what counts as harm inflicted by the market, together with the responsibilities that flow from the possibility of this harm and the entities susceptible to being ravaged by it. In other terms, we see emerge here an economy of attention that creates and reminds us about what it is that matters. It thus leaves aside or reduces the importance of other acts, other wrongs, and other vulnerabilities.

Accordingly, there are some sectors of activities and types of commodities, above all food and clothing products, that occupy the greatest share of the discourses on ethical consumption. Yet other sectors are only able to forge a path through the moral life sketched by ethical consumption with great difficulty, such as dematerialized technologies, which have the strange power of rendering invisible the material reality on which they rest—for example, the mineral components of smartphones (copper, lead, nickel, silver, tin, aluminium, and so on)—as well as the conditions of its exploitation, and its possible exhaustion. Similarly, few people have ventured the idea that music via streaming ought to be rejected because it damages the environment more than vinyl records, since it requires infinitely more energy (Devine 2019). What thus implicitly appears is an ethics that latches onto a "matter" able to be figured very concretely, a matter that signal a certain heaviness and has a rustic character. Correlatively, the labour that attracts concern is generally that involved in transforming such matter—work in which there is an especially heavy presence of bodies. So, for example, inasmuch as it is modelled on the notion of fair remuneration, the question of a fair price will overlook the universe of books and cultural goods.

In addition, the practice of responsible purchasing makes sense only in the framework of the official and legal market. In vain does one find a way to ethically consume banned—but nonetheless commonly consumed—goods, such as drugs. Procuring your stash of psychoactive substances can only be done with a disregard for the conditions of production, transport and business through which the goods become available. Upon closer inspection, this point contains an irony to the extent that ethical consumption presents itself precisely as a practice outside the institutional framework of the market. It is indeed because the legal and regulatory framework of commercial exchanges is seen as inadequate or ineffective that an individual ethics is established as necessary, an ethics that would outflank this framework by reintroducing as direct a relation as possible between producers and consumers through a collective organization of trade exchanges. By contrast, illegal practices, in which consumption already takes place outside formal market frameworks, seem barred from ethical behaviour.

In fair-price practices, where price is determined by a remuneration that attempts to do justice to the duration and strenuousness of the labour, the question that arises is that of work and its conditions. This question is also implicit in practices that aim to change the timing and rhythm of payments: accordingly, subscription systems have been designed in the case of peasant agriculture to address the rhythms that limit this work and the natural hazards that can reduce it to nothing. But these are only some of the important aspects of work and its modalities. Among the criteria that establish what counts as a fair choice, conditions that promote wage equity and democracy within the producing company scarcely feature at all.[3]

To consume ethically also means to buy that which would be otherwise consigned to the waste heap, that which is not new. This includes refraining from purchasing goods that will end up decaying into a series of simple waste products—packaging, objects, and components. Such concerns can then lead to an emphasis on trading products that can be produced on demand. Showing care here would thus mean to show a sense of measure. It would be a matter of limiting the volume of waste and the depletion of stocks of matter, or again to curb the immediate transformation of nature into scraps. Further, in the gesture of purchasing things that would be otherwise consigned to elimination, there is something else, which is, to take up a distinction made by François Dagognet, an attempt to transform waste into scoria: where waste is simply "what it is," a consequence of deterioration, scoria retain something good (*un meilleur*) that cannot be removed from them, at the same time as they exalt heterology" (1998: 64). In the piling up of used goods, the ethical consumer thus locates "a real abundance, the signs of a belonging to what one could call 'being'" (1998: 12).

By means of that with which it concerns itself, then, ethical consumption effectively creates a map of what the market does damage to, above all the body as that which is wounded through strenuously physical work, and matter in its rustical nature. It catalogues ways of doing damage—by exclusive and aggressive appropriation, by reducing scoria to waste, and so on. It even marks out the market that does the damage, which is to say the market that is legally regulated.

When the concern bears on nature, it is often preoccupied with the animal and its well-being (as pre-eminently defined by the space animals are granted and their access to the outdoors) (Norwood and Lusk 2011) and is thus foremost an attention to the physiological needs and suffering of familiar organisms. The buying of palm

[3] Axel Gosseries being an exception (2005).

oil-based foodstuffs has thus begun to present an ethical problem, above all because its production threatens the living environments of orangutans.

When it comes to human beings, the concern for others that is manifested in ethical consumption bears with insistence on women, especially to the extent that they are qualified as mothers or children from faraway places (the primary question here is thus that of their work), as well as on indigenous groups—specifically, the additional income that they could receive from getting a "fair" price is thus regularly reallocated to educating the producing community. The concern for others here thus focuses on entities, nature, women, children, and Indigenous peoples, which are all rather traditionally viewed as vulnerable but are also assigned a place in hierarchical dichotomies as the Other of power.

But all these possibilities merely amount to "offers" that the economic system provides, proposals that the market can make or not (and does not, for example, on the shady margins)—proposals that are served up. Limited to the carrying out of acts of refusal or acceptance of such proposals, the concern for others thus becomes a concern devoid of attentiveness, which takes the form of a qualm over the offers put forward, rather than a concern that is active and curious. Such concern is even mechanically produced by applications that quantify and rate various brands' respect for different ethical principles. By selecting the ethical principle that the consumer deems important, these applications evaluate businesses on identical rating scales and provide information to enable the decision-making of the consumer, who is thus unburdened of having to engage in any exercise of attention, or of an affective implementation of care.

2 Concern for Others Against Immoderation

Ethical consumption essentially strives to avoid abuse. It thus coincides with a self-imposed limitation on one's behaviour. Since capitalism draws on and exhausts (nature, labour power, and so on), correcting this tendency requires that one show good measure and holds off committing acts of unmitigated using up (of resources, vital energies, and so on).

Caught within this effort, the concern for others that underlies ethical consumption thus slips underneath the borders of the Marxist categories used to critique capitalism, categories that separate out production and reproduction, exploitation (of labour) and appropriation (of nature); where exploitation works to transfer value to the capital on the pretext of "free" and contractual exchange, appropriation itself rests on brute confiscation—of work, land, animals, minerals and energy, reproductive capacities, and organs. More precisely, ethical consumption grants the category of appropriation a particular and strong precedence. For example, when it comes to the development of some forms of mineral extraction, it sees in the resultant transformation of natural environments into sterile lands stripped of phreatic zones, a phenomenon no longer taken as collateral damage of market functioning, but as the heart of the process of accumulation. This sort of position converges with a more philosophical critique of contemporary capitalism that has recourse to concepts such as "expropriation," "appropriation," and "dispossession,"[4] where the aim is to establish a continuity between the confiscation of work, the confiscation of certain individual capacities, notably reproductive and sexual ones, the confiscation of non-human organic activities, and even the confiscation of organs or metabolic liquids.

Now, what ethical consumption sees in accumulation by appropriation is simply the voracity or predatory urge that subtends or accompanies it. Better still, it makes both coincide. The semantic field of predation is significant in writings on equitable trade, especially in works in English or in those that comment on English-language literature on this topic. In these texts, the similarly warlike vocabulary of "raid" is also often used, such as, for example, when attempting to describe the mainspring of appropriation as residing in the raids made by speculative funds and other institutions of financial capitalism.

This understanding of appropriation *as* an impulse, mood, or passion, derives, first and foremost, from the conviction lodged at the core of ethical consumption, according to which individual virtue—that which enjoins moderation—is the remedy to the nefarious effects of the market.

The idea of virtue is explicit in Aristotle's economic theory and was also a distinct part of the project of consumer cooperatives that burgeoned in Europe from the late nineteenth century to the mid-twentieth. These cooperatives sought to promote this idea to discourage their members from perpetuating practices that presupposed the certainty of a world at one's disposal, for example, practices of wastefulness, or demands for fresh bread in the morning, desisting from which would enable bakers to stop having to do night work. Charles Gide thus argued that cooperatives had an endogenous educational function, thanks to which the members would come to impose limitations on themselves. They would be the site of moral progress, the idea being that "in abolishing the pursuit of profit as the only real motive of economic activity—substituting for it the sole aim of satisfying needs—whilst abolishing advertisement, lying, cheating, and inducements to extravagance, cooperation will succeed in establishing the reign of truth and justice in exchange" (1921: 8).

Today the motif of virtue arises as a form of assessment of the intensity, sincerity and coherence of the effort made by individuals to moderate themselves. Ethical consumption calls for the sharpening of virtue, an acuity that demands self-discipline, temperance, restraint, and even frugality. Purchasing is then the site for an economy of remorse and self-contentment, which stems from an examination of the gap, or lack thereof, between the promised ethics and the reality of actions. The deployment of the required virtue is accompanied by a continuous and reciprocal observation, fed by social networks, of conduct and choices. It therefore implies that you open yourself up and maintain a culture of transparency. This imperative of transparency reminds us that the ethical content of an act of buying a packet of biscuits stemming from eco-responsible agriculture comes to nought if it is followed by an order of heavily packaged shampoo. Ethical consumption can only be such if

[4] See for example Harvey (2003) or Moore (2015).

it is a line of conduct, a continuous addition of non-contradictory acts, composing a mirror image of an appetite always liable to shirk restraint.

Indeed, as Philippa Foot has pointed out, virtues are indeed essentially "corrective," that is, the value of most of them (including temperance, with which we are concerned here, but also courage or charity) stems from a temptation that the point is to resist, or a weakness of motivation that is to be overcome (Foot 1978: 8).

To consume ethically amounts to controlling both an appetite for predation and an urge to make unfair choices, which are easier (as they require no effort of inquiry into a commodity's origin or conditions of production), more in line with our interests (as they are cheaper), and more in tune with our inclinations towards abundance and ostentation. It amounts to refusing the enjoyment provided by a world deemed to be at your disposal. It amounts to resisting the fever of acquisitiveness, whether individual or collective. Because our dispositions automatically lead us to pursue what we should be fleeing from, an ethical consumption demands that we have instilled in us a continual, built-in reminder of order and continence—a virtue.

For this ethics of consumption, appropriation is not so much a set of mechanisms for reducing production costs, which can rest on scientific revolutions or processes of localised devaluation, as a behaviour, or even a drive; with it, brutal exchange becomes an anthropological fact that is taken to another level by capitalism, or is even a pure creation of the latter. Appropriation becomes a general and drive-like attitude, a disposition to act according to principles of conduct that are not so much as formulated, but rather are the sign of our fulfilment as economic actors, or as actors *tout court*. This is the sense in which appropriation ought to be modified. It is as if capitalism presupposed a relation of ownership to things that coincided with the destruction of those things that was the simple result of the greed and predation that it urges. If it is possible to correct capitalism by stifling this type of appetite, then market morality is essentially a matter of goodwill, of victory over one's passions.

According to the grammar of ethical consumption, this mood of predation is open to all, small and big market players. But, by banding together in their exercise of restraint, to the small is imparted the power to thwart the appetite of the big—or perhaps to disarm this appetite by feeding it in a particular way, insofar as it is a question of forcing companies to modify their offers, by making them more "ethical" as they strive to maintain or increase their own profit margins.

As depicted by the critique of ethical consumption, capitalism is far from being simply a plunge into "the icy water of egoistic calculation," as per Marx's expression, but instead is based on a real passion, that of seizing, devouring, and destroying (Marx and Engels 2012). The mode of production is thus redefined by the psychology of individuals, rather than by its economic basis or the social relations that organise it. The whole systemic logic of capitalism is then left outside the scope of criticism. In particular, intersubjective relations are perpetually mediated by the market, which necessarily adjusts and de-adjusts them. This is particularly clear in what happens to concern for others when it is expressed in terms of price.

3 Concern and Price

Showing care on the market often means agreeing to pay a "fair" price. A fair price is one that is not simply determined by the meeting of supply and demand, but that integrates "the reality" of certain costs which tend not to be taken into account by the mere market: production costs, costs of converting to and respecting Fair Trade criteria (e.g. in terms of work organisation), costs incurred by the development of sustainability, environmental costs, or even a premium to improve the living conditions of producers (e.g. when it comes to contributing, through a purchase, to the development of a region of the world).

The work of Aristotle, and in particular Book V of the *Nicomachean Ethics*, is regularly mobilised by the proponents of fair trade; and whether it is or not, the reasoning it contains makes it possible to bring to light some of the underlying characteristics of the idea of a fair or just price.

With Aristotle, a just price can be conceived as the price that ensures the equality of the parties, or more precisely, the equality of the advantages that each derives from the exchange. Injustice thus consists of one party having too much and the other too little advantage, or, on the contrary, disadvantage. Restoring the equality of the parties to an exchange is an act that falls to the judge. The parties behave justly when they agree in the way a judge would ask them to do ("whenever the suffering is measured, the one is called a loss, the other a gain. As a result, the equal is a middle term with respect to what is more and less" (2011: 97)). The justice in question here is therefore corrective or restorative justice, which is achieved through an exchange, as opposed to distributive justice, which is achieved through sharing or distribution. It is immediately clear that in this design, there is no equality without *commensurability* ("all that is exchanged must somehow be capable of being compared" (2011: 100)). Correlatively, there is no equality possible without reference to money, for money is what institutes commensurability ("it measures everything—both excess and deficiency—so that it measures however many sandals are equal to a house or to food. Accordingly, as a house builder stands in relation to a shoemaker, so a given number of sandals must stand in relation to a house or food. For if this is not the case, then there will be no exchange or community" (2011: 100).

Aristotle's work thus opens the way to a long tradition, which accords commutative justice to the propriety of voluntary private transactions and makes the matter one of arithmetical proportionality. This tradition passes through Thomas Aquinas, according to whom in exchanges or "commutations something is rendered to a singular person because of a thing of his that has been received; this is especially clear in the case of buying and selling, where the notion of a commutation is found in the first instance. And so one must make a thing equal to a thing, with the result that to the extent that someone has more than what is his own from that which belongs to another, he restores to the latter what belongs to him" (1947: 410). Here again, a principle of absolute commensurability is applied. Any object can be exchanged for another, based on a quantified and reasonable comparison. The principle of a fair price makes exchange an art of measurement.

An art, not a science, despite the arithmetical referent. For it is important to stress that we are talking here about a balancing act that is first and foremost an ethical gesture. If justice, for Aristotle, is a mode of being ("that on the basis of which people are disposed to do just things" (2011: 90)), it is because it is a virtue or a quality of the mind; it cannot be confused with judgment, which is an activity of the mind. It is not primarily the judgement or evaluation that is just; justice lies in that which determines it as a condition, that is, the attention or concern for others. Justice is, therefore, a disposition, an orientation towards the other, towards what is advantageous to him: "This justice, then, is complete virtue, though not unqualifiedly but in relation to another person" (2011: 92). Thus the just price is not the product of a rational calculation, of an adequate quantification according to the criteria of science, of a scrupulous weighing up of different, well-understood interests, but of the exercise of a moral disposition.

Whether or not Aristotle is referred to, the current idea of a fair price bears on a concern for others that implies a reckoning, and attention to how individuals are caught up in the temporality of the market, a concern that is the mainspring of a corrective justice on and through the market.

A first consequence of this reasoning is that a fair price is by definition a simple ad hoc correction of the market determination of a good's value, the compensation of an imbalance that, it is postulated, can be absorbed, the curvature of a surplus that is blameworthy only through the exaggeration it betrays.

We see this clearly when the notion of a fair price is modelled on or made proportionate to the idea of a fair wage. For the latter is perfectly compatible with, or even requires, the idea that wage differences are legitimate. Marx carried out a severe critique of this in *Wages, Prices and Profit*, pointing out that, "Upon the basis of the wages system the value of labouring power is settled like that of every other commodity; and as different kinds of labouring power have different values, or require different quantities of labour for their production, they *must* fetch different prices in the labour market. To clamour for *equal or even equitable retribution* on the basis of the wages system is the same as to clamour for *freedom* on the basis of the slavery system" (Marx 1969: 46).

But more fundamentally, by having recourse to the idea of a fair price, ethical consumption contributes to a general logic of commensurability, which applies *in spite of and through* the ethical nature of the gesture, as the Aristotelian construct lays bare.

On the one hand, setting a fair price contributes to making countable what is not yet countable, particularly in the sense that perverse market effects are cut out and quantified. Take, for example, the establishment of a carbon price to promote renewables, whereby one takes into account the costs of damage caused by global warming ("externalities") as well as the interests of future generations. A difficulty here lies in determining what is to be included among all the externalities, as well as how to value in monetary terms the disappearance of a species, a landscape, or even a few thousand people without resources. Another difficulty resides in the monetary value that it would be desirable to give to the work of nature: how to value the work of pollination by bees or carbon sequestration by forests? Yet another relates to the assessment of costs for damages that will occur several decades or centuries from now, which also requires a mathematical cutting out of time.

But the essential matter here is not the accuracy of the attempt to funnel certain parts of reality into the world of numbers; it is simply the fact of their being so funnelled. The commensurability generated by the market stems above all from the power of money, which institutes market numbers by imposing itself as a standard: "money is what fully institutes market relations as counted relations" (Orléan 2015: 208). The market reduces things, nature, and human beings to a set of parameters that allows them to be valued and exchanged, which is at the same time a setting in equivalence. But general commensurability is achieved through the language of costs and prices even when the aim is not to "commodify" directly what is not yet fully commodified, such as nature. The monetisation of biological diversity is thus also conceived as a tool to assist public decision-making and as a way of giving a monetary value to ecosystems for heuristic purposes, in order to "make the invisible visible," so that economic actors become aware of the value of ecosystems, including for their own activity (Tordjman 2021: 209).

Similarly, an increasing number of operators are developing methodologies to establish a quantified relationship between the destruction of ecological environments and desirable compensation actions. This is achieved using a table of abstract functionalities to which a coefficient is attached—according to their recommendations, for example, a hectare with good quality soil can be exchanged for a larger area that is poorer in quality (Dauguet 2021). Beyond the technical device, and its weighted equivalence of heterogeneous functions, the assumption is that it is possible and moral to monetise and to equate our moral scruples. Not only does a graduated line order what it is more or less regrettable to destroy or neglect, but the idea of compensation absorbs that of reparation, in the development of a new trade in indulgences, where the effect is notably to posit that justice can be done through price, by a ratio between numerical quantities.

If to claim or establish a fair price, it is necessary to transform damage into a cost, a wrong into a subtraction, that is to say, it is necessary to count the world, to slice it into quantities, this also requires comparing and weighing it up, and doing so *through the mediation of the market*. On top of the active search for a mathematical formula or an encoded quantity that serves as a standard as such, the market exchange institutes measurement; "it is through barter that nonidentical individuals and performances become commensurable and identical" (Adorno 1973: 146). What is lost is the uniqueness, the singularity, of things, beings, environments, and situations. As Adorno and Horkheimer state in *The Dialectic of Reason*, borrowing from the Marxian analysis according to which bourgeois society is structured by the exchange relation and the commodity-form, "Bourgeois society is ruled by equivalence. It makes dissimilar things comparable by reducing them to abstract quantities" (2002: 4). The very idea of a fair price presupposes a metric by which everything is ordered. Out of a concern for justice and exemplarity, ethical consumption sets in motion a painstaking effort of calibration. It is therefore necessary to understand that the measure it uses is both the opposite of excess and the condition for exchange. In fact, a fair, appropriate measure presupposes comparison, and comparison, interchangeability. For comparison is based on the existence, or creation, of equivalences, as well as a principle of reduction to unity.

The effects of the equivalence form carried by the mediation of the market proceed even independently of the phenomena of transcription into pure costs. Rights around emissions trading reduce the quantities of the various greenhouse gases to their "CO2 equivalent," that is, to the volume of CO_2 that carries the same radiative power. This means that activities as different for the various parties of our form of life as rice cultivation (a source of methane), slash-and-burn agriculture (a source of non-fossil CO_2), or motor sport (a source of fossil CO_2), can be considered equivalent. Reducing all greenhouse gases to their common characteristic—their radiative power—is the condition for their exchangeability, although it is by no means indifferent, from the point of view of our form of life, whether we reduce rice production or put an end to car racing (Tanuro 2012: 113). Similarly, measures designed to improve the habitats of amphibians can come to compensate for the destruction of bat populations, as long as the countable relation is respected. In this way, capital creates a Nature in its own image, infinitely quantifiable and interchangeable, so that, as Jason W. Moore has pointed out, each piece is equivalent to another. It becomes a homogeneous quantum that can be divided and transformed at will. Corn is another striking example, as a "flexible crop" that provides the raw material for anything and everything: ethanol, food, building materials, industrial products, and so on. Quantity thus becomes the mode of existence of Nature, as of work or time. It evades the lived experience that human beings might have of these things; each time, it erases their specific contribution to a form of life.

In other words, the incommensurable is not only masked by generalised equivalence but the latter gets converted through the everyday into the order of normality and reason. In the very idea of a fair price, there is an objectively and compulsively realised equivalence, which repeats endlessly that a measure can be found in any exchange. And generalised commensurability rests on the distinction and combination of a dimension and countable motivations of exchange, and another motivation that claims to be uncountable, but which is no less intrinsic to the comparison of everything with everything. This is precisely the concern for others, human and non-human, that is at the heart of ethical consumption. It engenders their negation as singularities and makes them disappear by subsuming them into a fungible whole.

Thus, ethical consumption not only fails to master the mechanisms of capitalism that it targets; it actively contributes to perpetuating them and thus also the social totality that these mechanisms feed and gloss. Care for others as deployed on the market, being without any hold on the particular or individual, braced in an exercise of mastering capitalist hubris, indeed becomes turned inside out and, insofar as it is caught in the language of prices, is made an agent of generalised commensurability.

References

Adorno, Theodor W. 1973. *Negative dialectics*. London/New York: Routledge.

———. 2006. *Minima Moralia: Reflections on a mutilated life*. London: Verso.

Aristotle. 2011. *Aristotle's Nicomachean ethics*. Chicago: University of Chicago Press.

Dagognet, François. 1998. *Des détritus, des déchets, de l'abject: une philosophie écologique*. Paris: Les empêcheurs de penser en rond.

Dauguet, Benoît. 2021. *Mesures contre nature: Mythes et rouages de la compensation*. Caen: éditions Grevis.

Devine, Kyle. 2019. *Decomposed: The political ecology of music*. Cambridge: MIT Press.

Dubuisson-Quellier, Sophie. 2009. *La consommation engagée*. Paris: Presses de Sciences Po.

Ferrarese, Estelle. 2023. *Le marché de la vertu. Critique de la consommation éthique*. Paris: Vrin.

Foot, Philippa. 1978. *Virtues and vices and other essays in moral philosophy*. Berkeley/Oxford: University of California Press/Blackwell.

Gide, Charles. 1921. *Consumers' co-operative societies*. London/Manchester: T.F Unwin/The Co-operative Union.

Gosseries, Axel. 2005. Qu'est-ce qu'un consommateur juste? In *Ethique et commerce. Réalités et illusions*, ed. F. Mertz. Paris: L'Harmattan.

Harvey, David. 2003. *The new imperialism*. Oxford: Oxford University Press.

Horkheimer, Max, and Theodor W. Adorno. 2002. *The dialectic of enlightenment*. Trans. Edmund Jephcott. Stanford: Stanford University Press.

Marx, Karl. 1969. *Value, price and profit*. New York: International.

Marx, Karl, and Friedrich Engels. 2012. *The communist manifesto: A modern edition*. London: Verso.

Moore, Jason. 2015. *Capitalism in the web of life*. London/New York: Verso.

Norwood, Bailey, and Jayson Lusk. 2011. *Compassion by the pound: The economics of farm animal welfare*. Oxford: Oxford University Press.

Orléan, André. 2015. *L'empire de la valeur. Refonder l'économie*. Paris: Seuil.

Sassen, Saskia. 2014. *Expulsions: Brutality and complexity in the global economy*. Cambridge, MA: Harvard University Press.

Tanuro, Daniel. 2012. *L'impossible capitalisme vert*. Paris: La Découverte.

Thomas Aquinas. 1947. *Summa Theologica*. New York: Benziger Bros.

Tordjman, Hélène. 2021. *La croissance verte contre la nature*. Paris: La Découverte.

Chapter 9
Carelessness and Ignorance: The Epistemic Vices of the Privileged Analyzed Through Their Waste

Brunella Casalini

Abstract In this paper, I bring social feminist epistemology, ethics of care, and waste studies into conversation. I do so by reflecting on two alternative narratives of human evolution told by Ursula Le Guin in *The Carrier Bag Theory of Fiction*. Technology plays a central role in both. In the first story, the carrier bag's invention tells of humanity using what nature spontaneously produces and learning to preserve it to satisfy human needs. In the second story, the spear's invention tells of humanity having to sacrifice other species to survive; it follows a unidirectional, linear logic and cannot take the complexity of, or interdependence between, humans and the environment into account. The first narrative describes a caring and sharing humanity, one that pays special attention to the material practices that promote the flourishing of human and non-human life. Humans respect the principles of non-waste and responsibility for what is consumed. In contrast, the second narrative sees the prevalence of a dissipative human attitude. I shall argue that this latter attitude has led to today's acceptance of unlimited growth—to the willful ignorance of the corresponding and equally unlimited growth of waste. The world's rich and privileged countries are mainly responsible for the consumption of material goods but they have not yet come to terms with their waste. This is thanks to the possibility of relegating that waste to an 'Elsewhere' (often coinciding with the Global South). This "privileged irresponsibility" is based on technologies of distraction or deflection from the inequalities, social suffering, and injustices produced by the current global social infrastructure's "slow violence."

Keywords Waste · Carelessness · Feminist epistemology · Ethics of care · Discard studies

B. Casalini (✉)
Università di Firenze, Florence, Italy
e-mail: brunella.casalini@unifi.it

J. Braga (ed.), *Ethics of Care*, Advancing Global Bioethics 20,
https://doi.org/10.1007/978-3-031-73608-7_9

1 The Carrier Bag Theory: A Prologue

> It matters what matters we use to think other matters with; it matters what stories we tell to tell other stories with; it matters what knots knot knots with, what thoughts think thoughts, what descriptions describe descriptions, what ties tie ties. It matters what stories make worlds, what worlds make stories (Haraway 2016, p. 12).

In *The Carrier Bag Theory of Fiction* (1986), Ursula Le Guin tells a story that opens up a different world. This world contrasts with the one that has hitherto allowed us to imagine human evolution as beginning with the invention of the spear and the 'hunter-man,' who kills to survive and is the story's protagonist and hero.

Le Guin's alternative history follows in the footsteps of Elizabeth Fisher's anthropological work in *Women's Creation* (1979). Le Guin begins with the invention of a 'cultural device.' However, this time it is not a spear but a basket—a container for storing what is gathered from nature. A humanity that builds containers to store what can then be shared stands in opposition to one that develops via destruction and the invention of aggressive devices. As Le Guin (2019, p. 32) points out, both the basket and the home (and, we might add, even the city) play a containing role. We can also think of this as Donald Winnicott's "holding" (cf. Honig 2017). The basket and the home (and even the city) hold what is fundamental to the preservation of memory—to the remembrance of what is valuable.

Le Guin explains that the story itself is a "bag." It contains words—it carries images that continue to speak to us and convey an idea of the world: "A book holds words. Words hold things. They bear meanings. A story is a medicine bundle, holding things in a particular, powerful relationship to one another and to us" (Le Guin 2019, p. 34). Le Guin's musings about the bag make us ponder how we can belong to humanity. Can we do so in ways that do not involve prioritizing the value of the hunter-man and his pursuant toxic history?

Le Guin's narrative is especially powerful in our current time of crisis. Indeed, capitalist modernity continues at unprecedented speed along the devastating path forged by the primitive and "murderous" male. In Le Guin's words, technology and science continue to be seen as "weapon[s] of domination," with the "linear, progressive, Time's-(killing)-arrow mode of the Techno-Heroic" (2019, p. 36). The techno-heroic attitude leads humanity down a fraught path. This path involves dissipating the limited resources that a finite planet has made available to life. Technology is decoupled from the needs of life and bent to the purposes of human domination—a domination that must continually be reaffirmed, ignoring the costs to nature. The speed of innovation, growth, and destruction seem to somehow accelerate together. As Mark Zuckerberg asserted in 2009, "Move fast and break things. Unless you are breaking some stuff you are not moving fast enough" (quoted in Vinsel, Russell 2020, p. 6).

2 "The Waste We Want to Forget"

According to World Bank data, 2.01 billion tonnes of municipal solid waste alone is produced annually—a figure that is set to rise to 3.40 billion tonnes by 2050 (Kaza et al. 2018). In a race toward unlimited economic growth, humanity's destructive eagerness manifests itself in the rampant production of waste. Like an ominous shadow, this production necessarily follows humanity's increasing consumption, the acceleration of extractive activities, and the breakneck industrial manufacturing of goods destined for obsolescence.

This trend is dictated by impetuous fashion and rapidly changing technologies. However, disuse and discarding are often intentionally planned right into product designs (Latouche 2012). All of this is creating a prospective "loss of the world" (as Arendt [1958] would have put it). Indeed, the stable and lasting public spaces where people act communally are disappearing. The individual is reduced to mere *animal laborans* or *homo consumens*. The human species is, thereby, putting itself and most of its earthly cohabitants in mortal peril.

The dangers of losing the "world" and destroying the "earth" (to use Arendt's lexicon again) are tightly linked. The threat comes from the logic that global capitalism rests on. This is the economic reasoning that Val Plumwood (2001) described 20 years ago as "arrogant, sadistic, and indifferent." This reasoning fails to recognize the implications of our corporeal being and our material dependence on the environment. Such intentional denial and strategic ignorance generate cognitive and affective distortions. These distortions, in turn, generate a sense of irresponsibility, indifference, and carelessness, especially among those living in privileged contexts. Plumwood writes:

> We can find out what is available in the market to meet our desires and at what price, but we cannot without great difficulty trace where it came from or what its real costs are, in terms of the earth. Under current rules for global trade, it may soon be illegal to inquire. [...] commodity culture actively erases such connections, encouraging us 'to focus on the meanings we make, but not on our complicity in the economic networks through which people convert nature and human labour into the stuff and sustenance of everyday lives'. Remoteness negates responsibility, for consumers, workers and shareholders. In rationalist commodity culture, we are actively prevented from exercising care and living in ecologically-embedded and responsible ways (2001, p. 16).

Willful ignorance about the chain of events that brings food to the tables and goods to the homes of those living in wealthy countries is kept firmly in place. Even the almost obsessive preoccupation we are asked to have with household waste's production and disposal seems designed to keep things as they are (I shall return to this issue below). As Kate Crawford writes, "Ignorance of the supply chain is baked into capitalism [...]: the left hand cannot know what the right hand is doing, which requires increasingly lavish, baroque and complex forms of distancing" (2021, p. 35). Moreover,

> China supplies 95% percent of the rare earth minerals. China's market domination, as the writer Tim Maughan observes, owes far less to geology than to the country's willingness to take on the environmental damage of extraction. Although rare earth minerals like

> neodymium and cerium are relatively common, making them usable requires the hazardous process of dissolving them in large volumes of sulfuric and nitric acid. These acid baths yield reservoirs of poisonous waste that fill the dead lake Baotou. This is just one of the places that are brimming with what environmental studies scholar Myra Hird calls 'the waste we want to forget' (Crawford 2021, p. 36).

Gabrielle Hecht (2018, p. 111) has defined the "Anthropocene" as the "apotheosis of waste," while Mario Armiero (2021) speaks of the "Wasteocene." But, what is our relationship with waste? The way we take care and dispose of waste suffers from obvious distortions. The rhetoric of waste differentiation and the circular economy that governments, corporations, and recycling companies use shifts responsibility away from the production process and delegates it to individuals and families. This choice to focus on what happens after consumption avoids placing the production process or the choice of materials in the spotlight. At the same time, it prevents consumers from demanding objects with guaranteed durability and the right to repair (see Hird 2022; Liboiron and Lepawsky 2022).

We face a complex reality where waste production is removed from our gaze and located in an Elsewhere, which often coincides with previously colonized areas. Dealing with this situation means addressing the lesser-known and well-hidden side of the capitalist system. This is its reliance on accumulation strategies—strategies that do not only rest on exploiting free labor but also operate by pushing millions of people into informal, gray economies. These strategies do not just expropriate and despoil. They can also transform territories into veritable dumpsites by (a) outsourcing the most polluting industries and (b) exporting first-world waste to Global South countries and regions inhabited by indigenous peoples (Liboiron 2021).

As reported in the literature on care and social reproduction, the most disadvantaged social groups often clean spaces and care for the bodies of the elderly, minors, and the disabled in the rich West (see Vergès 2019; Prattes 2020). At the same time, they are called on to absorb the waste from Western production and consumption systems, which is causing serious damage. This is not taken into account when calculating production costs. The richest portion of the world's population is ensured prosperity through destruction—through the "slow violence" (Nixon 2013) of pollution processes. This is destroying the environmental conditions that sustain life in the world's poorest countries. It is due to an almost invisible global social infrastructure, one that enables "rich, white, settler-colonial, heteronormative reproduction" (Murphy 2018, p. 109) while tacitly disabling other populations.

3 Carelessness, Distraction and Ignorance

Social epistemologists, agnotologists, and those working on the so-called epistemology of ignorance study both how knowledge is constructed and how ignorance (or non-knowledge) is produced (Proctor and Schiebinger 2008; Sullivan and Tuana 2007; Medina 2013). In what follows, I shall employ feminist social epistemology and, more specifically, the epistemology of ignorance to analyze certain political *techniques of distraction* (Alexander and O'Hare 2020). These are the techniques that lead us to accept a conventional boundary between what is and is not worthy of care and attention. I focus here on waste management and the construction of an imaginary boundary between (a) what is deemed worthy of value and care and (b) what is considered waste. I contend that such a construction is linked to salient power strategies (of the Foucauldian sort). These operate by creating a reality—by making certain things appear natural, good, and true (cf. Liboiron and Lepawsky 2022, p. 7). Certain things are deemed worthy of interest and attention while others are neglected or rendered invisible.

Before proceeding with my thesis, I would like to discuss the notion of privilege and the epistemic vices of those with systemic social advantages. Understanding how privilege works allows us to better understand the difficulty involved in implementing cognitive processes that are up to the challenges we are facing. A difficulty stems from the fact that responsibility for what is happening cannot be attributed to a generic "we"—to humanity in general. As those who call the current era the "Capitalocene" (rather than the "Anthropocene") point out, the responsibility lies with global capitalism and its internal logic (Malm 2016; Moore 2016).

The term "Anthropocene" might seem neutral and universal, but it risks invisibilizing the relationship between capitalism, colonialism, and racism (cf. Armiero 2021, ch. I). This occurs through the privileging of a Eurocentric narrative. Yet, silencing, invisibilizing, and failing to listen to those most directly affected by the Capitalocene has notable negative consequences. These are the people who live on the margins of rich societies and in the world's poorest countries. Normalizing the suffering that slow violence produces—the violence inherent in routine affairs of the unequal social structures our societies rests on—ensures that the privileged can maintain their illusion that everything will continue according to the rules of infinite growth. But, the costs mainly fall on those whose lives are deemed expendable and on future generations. This normalization impedes the rich from understanding and knowing reality; it severely hinders their ability to learn from what is happening. This continues to be the case even when we are faced with imminent catastrophe. Unable to understand the consequences of their actions and the web of interdependence in which we live, the privileged fail to develop a sense of "response-ability" (Haraway 2016). The term 'response-ability' is used in contemporary feminist science studies to articulate relational ethics and politics dealing with the complex practical relationships we have with both the human and non-human world (Liboiron 2021, p. 17).

As the critical theory of care maintains (cf. Fitzgerald 2022), the privileged should cultivate both (a) an awareness of their vulnerability and (b) a horizontal gaze, one that is attentive to the dimension of relationships (the network of interconnections and interdependencies their actions give rise to). Doing so will reinforce practices that live up to the concrete capacity of providing responses aimed at repairing and maintaining the world. This should allow everyone to live as well as possible.

The vulnerability I have mentioned is, first and foremost, a kind of epistemic vulnerability. This relates to the rich failing to grasp the full extent of their actions and the consequences arising therefrom. They should, then, listen to the voices of those who suffer the indirect consequences of their choices. Correct knowledge about the problems that the Capitalocene raises becomes all the more serious when we consider the following: If the privileged (understood as affluent inhabitants of the richest countries) have an interest in resting on, rather than questioning, their privileges, then those who hold economic and political power have an even stronger interest in maintaining what agnotologists have called "strategic ignorance" (Proctor and Schiebinger 2008) or "manufacturing of doubt" (Oreskes and Conway 2011), the deliberate production of ignorance.

4 The "Invisible, Weightless Knapsack" of Privilege

Le Guin's carrier bag theory speaks of a humanity whose main activity consists of the following:

> [Put what you want, because] it's useful, edible or beautiful, into a bag, or a basket, or a bit of rolled bark or in a leaf, or a net woven of your own hair, or what have you, and then take it home with you, home being another, larger kind of pouch or bag, a container for people; and then later on you take it out and eat it or share it or store it up for the winter [...] (2019, p. 32).

The carrier bag theory tells the story of humans who, knowing that they will have to carry the weight of the bag, select what is essential—what is beautiful, sacred, or affectively important. These things' value is recognizable and socially acknowledged; it is worth taking care of each day—worth maintaining.

A "humanity of care" carries a heavy bag of essentials, whose value it recognizes. People try to save, even in the face of the worst catastrophes. This picture is more likely to be represented by women, the oppressed, and the most disadvantaged social groups—by those who struggle to meet their essential needs or the needs of those who cannot provide for themselves.

It is, perhaps, no coincidence that Peggy MacIntosh describes the opposite reality, one of white privilege. She uses the metaphor of an "invisible, weightless knapsack," which contains a set of "special devices, insurance, tools, maps, guides, ciphers, passports, visas, clothes, compass, emergency equipment and blank cheques" (MacIntosh 2020, pp. 29–34). These are resources that can help those possessing them to orient themselves in the world, allowing them to be at ease in most everyday situations—in everyday life.

The privileged carry an invisible knapsack because they can afford to be unaware of the fact that they are wearing it and that it has been filled with essential social assets. Indeed, their privilege consists in acting as if their advantage is earned (or deserved). They can ignore the fact that their privilege results from a system designed to keep them in a dominant position. The knapsack McIntosh speaks about is, therefore, not weightless because it is empty but because privilege produces both (a) amnesia (those belonging to the dominant group can forget that they have it and how they managed to get it) and (b) "anaesthesia" (the privileged might not feel the weight of wearing it—the heaviness of patriarchy, whiteness, and heteronormativity) (Bailey 2021, p. 87). They might not sense that an unequal regime has allowed their knapsack to be filled with essential resources. Indeed, it has become a kind of trump card.

Privilege does not only affect the cognitive level but also the affective level (Molinier 2018, p. 60). It does not only produce selective memory and knowledge, which allows the reality of oppressed groups to be ignored. It is also at the origin of what Joan Tronto calls "privileged irresponsibility" (1990, 1993, 2013; see also Zembylas et al. 2014; Pease 2020). Privileged irresponsibility derives from indifference, carelessness, and emotional distance from the situation those living in conditions of oppression find themselves in.

In short, the privileged have the privilege of thinking that the plight of the oppressed does not concern them. This represents a kind of "psychological posture," which—as Pascale Molinier explains—

> rests on a complex device—deological but also of segregation—that prevents the privileged from thinking about dimensions of reality that could cause them embarrassment or prevent them from enjoying their privileges or harm their sense of justice (ibid.).

As such, the knapsack's weightlessness, on the one hand, refers to the fact that those who belong to privileged groups have done nothing to gain the advantages available to them. On the other hand, it refers to the fact that the wearer might be unaware of the cost they inflict on other social groups. They might be indifferent—affectively anesthetized—to the costs their privilege entails for others.

Sara Ahmed (2017) has expressed ideas that are close to MacIntosh's, but invoke a different image. Ahmed defines social privilege as "an energy-saving device" (2017, p. 125) because not everyone has to make the same effort and spend the same amount of time reaching equivalent goals. For some bodies, even just standing still—surviving—is exhausting. Privilege offers a support system—a kind of "buffer zone" (Ahmed 2017, p. 148, p. 219, pp. 237–238). It cushions the impact of any falls, and can, thereby, "reduce the cost of vulnerability" (Ahmed 2017, p. 238). If something happens—if you fall—then "you are more likely to be looked after" (ibid).

5 The Epistemic Vices of the Privileged

MacIntosh only briefly mentions the epistemic vices of the privileged in his analysis. It is, though, the subject of careful analysis in José Medina's *The Epistemology of Resistance*. Medina defines 'epistemic vices' as "structural and systematic" defects. These are rooted in cognitive attitudes and personality structures that condition one's relationship with both oneself and others. They prevent one from correcting oneself and learning from contact with 'the Other'. Medina speaks of privileged people's "active ignorance," which can give rise to various epistemic vices, including "epistemic arrogance."

In its extreme form, epistemic arrogance causes a subject to perceive reality as their own creation—as totally under their control. "Epistemic laziness," in turn, represents a lack of curiosity, which grants privileged people the possibility and the comfort of remaining in a state of ignorance (Medina 2013, pp. 30–40). Medina writes:

> The epistemic vice of laziness converges with that of arrogance in seriously compromising one's learning potential and contributions to knowledge. Like epistemic arrogance, epistemic laziness also has a negative impact not only on the subject's perspective, but also on the cognitive perspectives of those around the subject and on the social knowledge that becomes available or unavailable to the relevant communities (2013, p. 34).

These two vices are mutually reinforcing. They contribute to the creation of negative epistemic injustices for society as a whole (cf. ibid.). They can lead one to overlook the consequences of one's actions and choices, especially when they affect those who are far away. Indeed, remoteness—whether spatial, communicative, temporal, or otherwise—is one of the elements producing distortions in cognitive and affective processes (Plumwood 2001). It thus influences decisions and makes the resulting actions ecologically irresponsible and even irrational.

The ignorance of the powerful—their carelessness and arrogance—is sustained and maintained in several ways. One of these is the systematic devaluation of the Other's knowledge and voice. We can trace this notion back to the phenomenon Miranda Fricker (2007) calls "testimonial injustice" (i.e., undermining and sabotaging oppressed groups' credibility, authority, and ability to reason). This is a powerful expedient for naturalizing a system of oppression and wavering responsibility for the injustices it rests on.

Mass media often puts these mechanisms into action when portraying the conditions of the world's poorest people. These people are sometimes shown working as waste pickers amidst mountains of rubbish. The media rarely question where the rubbish comes from, and even more rarely recognize that it comes from the West. So, we end up associating their existence surrounded by waste with a confirmation that people from the Global South are uncivilized. Some might even view sending waste to poor countries as an act of generosity on the part of rich countries. For example, exporting second-hand computers (actually e-waste) to countries such as Ghana might be justified as a charitable donation. It might be viewed as an act that (a) gives poor young people a job opportunity and (b) grants Africans the chance to access the Internet and thereby the digital innovation they would otherwise be excluded from given the costs involved. (See Perkin [2017] for more on the use of opportunistic discourses to justify the arrival of e-waste in Ghana).

6 Privileges and Boundaries

One of the traditional ways in which the privileged defend their position of systematic advantage is by constructing boundaries—dividing lines between opposites placed in hierarchized dualisms. Following Plumwood again, we can associate modern rationalism (and its negative outcomes) with a logic that operates on reality through categorizations and binaries (nature/culture, body/mind, dirty/clean, male/female, matter/spirit, private/public, etc.). These produce a specific idea of social order and, with it, a specific hierarchy of values. The devaluation of care is incomprehensible outside a devaluation of everything that is associated with nature, the body, the private, the needy, and the dirty. When we devalorize care, we are also devalorizing (a) the cycles of nature and (b) the daily repetition of the ordinary activities that form and transform. From this point of view, the social order is built to be inhabited by rational, sovereign, independent, and pure subjects. These subjects are heedless of the long-term and far-reaching consequences of their actions—actions that must be performed on an Other and an Elsewhere.

The dominant class of white males has freed themselves from the burden of care work by assigning responsibility to women and people of color. This practice is facilitated by (a) the boundary traditionally drawn between the public and private spheres and (b) the reduction of the private to natural affections and needs. This disengagement from the range of arduous and monotonous daily activities involved in caring for minors, the elderly, and the needy has freed white men's energies for other activities. Men do not have to recognize their dependence on these activities. They do not have to acknowledge the caring and affective work that their daily existence depends on. Thanks to the "tradition of separate spheres"—the separation of the public and the private—men need not learn about the effort and fatigue required to carry out care work. They need not recognize its value for social reproduction, and this ignorance allows them to continue imagining themselves as independent and self-sufficient individuals—to maintain the "autonomy myth" (Fineman 2005).

This strategy of separation—a separation between center and periphery—has also informed how the richest parts of the world have handled the issue of waste, especially toxic waste and the most polluting industrial activities. The burden is generally delegated to the most vulnerable populations. Doing so relies on *externalization*—the existence of sacrificial zones. In peripheral and invisible spaces, natural resources and a labor force of mostly informal, underpaid workers continue to be exploited. There is also an unloading of social reproduction costs related to the destruction of the environment and essential resources like air, water, and land.

7 The 'Dirty', the Waste, and the Othering of the Capitalocene

In modernity, the labor of social reproduction has been devalued through invisibilization—by disappearing within the four walls of the home. That said, its exploitation has not been restricted to the private sphere. In an important interdisciplinary field—discard and waste studies (cf. Gille and Lepawsky 2021; Liboiron and Lepawsky 2022)—the rationale for assigning cleaning and care tasks to women and racialized subjects is best understood by returning to Mary Douglas' discussion about the relationship between dirt and disorder.

In *Purity and Danger*, Douglas describes the social order as built on a system of dumping, discarding, and/or neutralizing the impure elements that might endanger it: "Reflection on dirt involves reflection on the relation of order to disorder, being to non-being, form to formlessness" (1966, p. 6). For Douglas, there is no such thing as "dirt" in itself. Dirt's identification rests on a process of classification—a process of putting it into form—of placing it in a socially meaningful and shared order. Douglas is especially interested in showing how power operates through the creation of clear boundaries around what represents cleanliness and order (1966, p. 32; see also Gille and Lepawsky 2021, p. 45).

Care, cleanliness, and maintenance are constitutive elements of the social order. They are part of the process of attributing economic, affective, religious, aesthetic, or utility-centered value. This recalls the way that neglect is reserved for what is expelled or discarded—for what is now considered worthless. The expulsion of (human and non-human) waste—of dirt and garbage—relies on (a) the existence of an Elsewhere where it can disappear and (b) an Other who takes care of its management. Like that which is expelled when constructing the social order, the Others who manage the waste and the Elsewhere that should make it invisible are considered unworthy of attention. On the one hand, this means that neglect, inattention, and indifference are just as essential as care, attention, and concern in explaining how the social order works. On the other hand, it means that practices aimed at creating order simultaneously produce multiple forms of inequality. We can say that the very distinction between (a) what is worthy of care and (b) what can and should be ignored and remain uncared for is constitutive of what is considered to belong to the political, in the sense of being considered worthy of political consideration and deliberation. Such a scenario gives rise to struggles over the boundary between care and neglect.

If every system produces waste, the form that waste has taken in modernity has unprecedented features. This may be because the search for purity—for purified categories and sharp boundaries—has become a form of obsession (as Hawkins [2006, p. 10] has noted, following Latour [1993]). It might be because of contemporary waste's unprecedented quantity, planetary diffusion, durability, and toxicity. Atomic waste toxicity, for instance, can last up to 100,000 years (i.e. for 3000 human generations) (a telling example of what Timothy Morton [2013] calls "hyperobjects"). The ways in which waste's existence has been removed from sight are also unprecedented. It is relegated to the Elsewhere—to the colonial Elsewhere (Hecht 2018, p. 111).

As mentioned, these spaces are invisibilized and the populations living in them are silenced. Their life prospects are damaged by a rich world—a world that benefits from the current capitalist economic system. That said, the blame does not rest solely on privileged consumers' desire to safeguard their standards of living. We should not forget how large international corporations artfully and deliberately create ignorance (with the help of complicit governments). This ignorance relates to the effects of production processes and the impact the Anthropocene has on certain populations and environments (just think of the harmful effects of smoking and climate change) (cf. Proctor and Schiebinger 2008; Oreskes and Conway 2011).

8 From the "Leonia" of Calvino to Today's Recycling Culture

A tale like the one told by Calvino in *Invisible Cities* appears only partly capable of portraying the reality we live in (even through imagination). Calvino depicts the consumerist irrationality that afflicts modernity through a description of the city of Leonia. The inhabitants of this city love innovations: "Leonia refashions itself every day." (Calvino 2010, p. 102). Its citizens' desire for novelty is so limitless that it can never be satisfied. Every morning they buy new objects, which are ready and waiting for the 'garbage truck' the next morning. Calvino wonders if "Leonia's true passion is really, as they say, the enjoyment of new and different things, and not, instead, the joy of expelling, discarding, cleansing itself of a recurrent impurity" (ibidem).

The citizens of Leonia's true passion seems to be removing dirt from themselves so that they never need to think about it again, even though it is piling up outside the city walls. It is hard to imagine that they really cannot perceive its existence, especially when the wind picks up and carries the stench of garbage into the city. To maintain the privilege of enjoying new goods every day, they are willing to ignore the mountain that grows a stone's throw from the city limits. If they truly recognized it, then they would be horrified. They would tremble at the idea of the dangers it hides. But, as Zygmunt Bauman (2003) notes, the citizens might be so disconcerted by the "simple truth" that they would fail to take in: "[T]he odious heaps of waste can only *not be* if they have not been made to be (by them, Leonians!) in the first place" (Bauman 2003, Kindle, Introduction).

By renewing itself every day, the city preserves all of itself in its defining form. What Leonia is—her identity—is defined above all by her waste. These are the things she throws away and wants to know nothing about, even though more and more objects are made of materials whose lifespan might outlast both the city and its inhabitants. As in Thorstein Veblen's *The Theory of Leisure Class* (1899), associating waste with the concept of conspicuous consumption means that (despite its intriguing narrative) Calvino's account captures only part of contemporary society's "reality of discard."

The waste from today's cities (both large megacities and small Western towns) continues to tell of its inhabitants' inexhaustible passion for *the new*. They are driven to unlimited consumption by both fashion obsolescence and technological obsolescence. But, above all, they are driven by planned obsolescence (Latouche 2012), which denies the consumer the right to repair. As Alexander and O'Hare (2020) observe, this situation builds "a future of calculated waste [...] embedded in the very design" (Alexander and O'Hare 2020, p. 433) of the product.

Unlike Leonia's inhabitants, some citizens of the privileged West are becoming aware of their responsibility when it comes to waste production—an awareness that is expressed in concerns for domestic garbage sorting and recycling. However, this narrow focus on waste from consumption appears to be another way of closing one's eyes to the real problem: waste in the modern capitalist system. Ignorance about this danger is preventing waste from becoming the object of adequate care and attention. What contemporary social epistemologists call "strategic ignorance" (Proctor and Schiebinger 2008) also comes into play here because focusing solely on domestic trash draws attention away from industrial waste.

The rhetoric of "blaming the consumer" can be seen as a further mechanism of diversion. It diverts our gaze with the aim of once again keeping the system functioning as usual (Hird 2022). If the waste problem is constructed with municipal solid waste in mind (i.e. the waste produced by consumption), then it seems natural to think that private citizens are primarily responsible for separate collections aimed at recycling and that local authorities should manage landfills. This narration and construction of the problem amplifies the responsibility of consumers and local authorities. The problem is that it conceals the waste produced by modern capitalism. This, in turn, prevents the emergence of an informed public, one that is aware and capable of demanding truly effective responses. As a result, the costs of waste management are not borne by those who are actually responsible for it, namely the companies.

9 The Limits of the Current Waste Narrative

The first distortion of information that the current narrative produces emerges from analyzing data collected on the total amount of waste produced in countries such as Canada and the United States. We can imagine that the situation in other industrialized Western countries will be similar. In Canada, for example, municipal solid waste constitutes only 2.6% of total waste generated when one factors in waste from mining, processing extracted resources, and industrial processes. This would be an even smaller percentage (around 1%) if one included waste generated by the military sector (Hird 2022). As Myra Hird notes, this means that modern capitalism's waste issue arises before the consumer even enters the scene:

> Our global waste problem is produced by the extractive, manufacturing and retail industries that have deliberately amplified the consumer's role in waste production in order to deflect responsibility from themselves and on to consumers (2022, Kindle).

The second limitation of the current waste narrative relates to the efficacy of recycling policies (which are far removed from the ideas and objectives of the circular economy). In the case of plastics, for instance, we must bear in mind that only a very small part of the waste that ends up in the recycling bins can actually be recycled. Its regeneration also often requires high costs when it comes to the energy needed to actuate the industrial process. These costs are higher than those involved in using virgin resources. Regenerated plastics are also generally of poorer quality than new ones.

Most of the plastic produced and used ends up in incinerators or landfills. In contact with the weather and sewage, it can leach toxic substances into the environment, polluting groundwater and soil. If not, it ends up in rivers, seas, and oceans or is transported to distant countries. China was the main destination until 2017, when it imposed a ban on the import of solid waste, sending the system into a tailspin. Today, countries in Southeast Asia and Africa mostly receive the rubbish. The most polluting mining and industrial processes have also been exported there (e.g. uranium mining procedures in Africa and textile and clothing production in Southeast Asia).

The waste that industrial processes produce is often invisible to the naked eye. It takes the form of tiny particles, such as dioxins, lead, mercury, pesticides, organophosphates, and the like. It consists of chemicals that leach into the environment, penetrating soil, water, air, and bodies. Plastics—mostly made up of tiny residues—have invaded seas and oceans and are now also found in living organisms. It is being scattered everywhere, but its presence is difficult to communicate to the public. People are used to worrying only when they see large piles of plastic bottles, bags, and similar waste.

Populations living near industries or polluting sites cannot escape the effects of these "chemical infrastructures." Such chemicals act on the endocrine system with metabolic effects that are only now being investigated in a new field of study, exposomics (Murphy 2018). For these populations, the daily reproduction of life is being compromised, often irreparably. This is due to the indifference of their governments and the carelessness of countries exporting waste and polluting industrial operations.

10 The Need for an Ecology of Attention

Piles of trash still leave Leonias. But, now it gets sent to other cities or states where people live by recovering and recycling the waste of the rich. Cairo's Manshiyat Naser neighborhood (also known as Garbage City) contains the Coptic minority—an ill-represented and marginalized group. For years, they have subsisted by means of a survival economy centered around recovering materials and debris expelled from the megalopolis. This represents one of the most effective and efficient recycling systems in the world (cf. Art Documentary 2021). At the same time, the Copts are creating their own identity and dignity by giving new life to waste through an informal economy of reuse and recycling. This serves as a kind of resistance against a system that tries to reduce them to "wasted lives" (Bauman 2003).

The circular economy intimates at the possibility of seeing all forms of waste disappear once and for all in an infinite, lossless production cycle of transforming waste into resources. Some might wonder whether this image can also serve as a way of inhibiting the only truly decisive drives: reducing consumption and taking responsibility for the production and destination of new products. As Alexander and O'Hare put it, the awareness that "there will always be waste that we will never know how to transmute into something harmless or positive is the first step towards a collective responsibility towards resource extraction and consumption" (2020, p. 16). Only attentive looks at global social structures—at the network of interconnections sustaining the capitalist system's waste management—can help us redistribute responsibility for the *care* of rubbish. Only then we can take responsibility for the evils, violence, and social injustices that have been created by the injudicious ideal of unlimited growth and accumulation.

To mature this outlook we need an "ecology of attention" (Citton 2021/2014), which eliminates the distortions that modern rationalism produces. These distortions move us away from practices of care and have been dangerously exacerbated by both the logic of neoliberal capitalism (founded on myths of meritocracy, measurement, and competition [cf. Lynch 2022, ch. 7]) and the current "attention economy." This economic regime exploits and manipulates our attention through the capitalist activities of corporations such as Google, YouTube, and Facebook. Georg Franck (2005) calls this "mental capitalism," which measures inequality in terms of where we place our attention. It focuses on who we direct our gaze at, whose presence we emphasize, and whose needs and speech we accord consideration. Attention is a scarce resource in this economic system. Inequalities grow between (a) those who consistently pay attention but receive none in return and (b) those who appear in the media and are thereby constant objects of attention (ibid.).

Thus, a new conflict takes shape, one that affects both poor and rich people in both poor and rich countries. The rich export large amounts of information and import large amounts of attention, while the poor attract only modest attention for the cultural products they manage to create (Citton 2021/2014, pp. 112–114). Attention is further subjected to the logic of profit. In many ways, this represents an unprecedented danger. The seriousness of this becomes apparent when we realize that what we pay attention to is also what we end up attributing value to; "our attentional processes are inextricably linked to our valorization processes" (Citton 2021/2014, p. 51, my translation).

Isabelle Stengers warns against this danger when she observes that, in an age where our attention seems besieged from every angle, we are strangely asked to pay no attention to where we are going:

> [W]hen talking about 'development' or 'growth' we have been asked not to pay attention. [...] What we have been led to forget is not so much the ability to pay attention, but rather the art of paying attention. If there is art and not just skill, it is because it is a matter of learning and cultivating attention – literally, of paying attention. I speak of 'doing' because attention, here, does not refer to aspects of the world that a priori would be worthy of it, but forces one to imagine, to monitor, to consider consequences that bring into play connections between things we are used to dealing with separately (Stengers 2013/2009, p. 51; my translation).

11 Closing Remarks

Attention is the incipit of any caring activity. It is a small "theoretical gesture" (Baraitser 2017, p. 14) lying behind what the good life should consist in—what we should value to live well in this world—what cannot be renounced. As Stengers (2013) reminds us, we must continue to keep attention in our bag after having duly considered the multiple interconnections between things. These interconnections include actions whose consequences fall on third parties but that we must recognize as our own responsibility. To avoid the evils that can result, the privileged should renounce their intentional indifference to the fate of the most disadvantaged. Multinational corporations, which today protect their interests through strategic ignorance and the creation of doubt by spreading disinformation, should be held accountable for their responsibility in climate change and pay for the damage caused.

In "Slow Violence and Toxic Geographies: 'Out of Sight' to Whom?" (2022), Thom Davies points out that we are not lacking sources, images, documentaries, testimonies, or stories that can teach us how to reason around our interdependence with other humans and non-humans (as Rob Nixon [2013] would have it). The web provides ample documentation for those who have the curiosity and wherewithal to search. Nonetheless, attention is still too easily diverted toward what is reassuring and offers immediate gratification. Those who hold the economy's reins readily take advantage of this desire to be distracted in defending their interests.

Natural disasters and environmental catastrophes are increasingly frequent, and they even touch the privileged. These are warning signs that can help awaken a collective concern and sense of responsibility. Environmental activism and commitments indicate a strong emotional need for action among those who see danger in the future. This is especially the case among the youth. There is a new, emerging global ecological consciousness, well aware of the interdependence of all living organisms. New generations are recognizing how the slow violence of environmental degradation—delayed in time and space—creates social injustice. It threatens the very conditions for reproduction and maintenance of life on the planet. The most disadvantaged social groups are called on to bear the waste from wealthy country's production and consumption, resulting in serious damage to the former's lives.

Political decisions and measures must force the economic system to consider the durability of objects, their real usefulness, their maintenance and repair, and their impact on our lives and the life of the planet. This applies throughout the non-human lifespan of those objects' components and their remains (Campbell and Wilson 2017; Vinsel and Russell 2020; Denis and Pontille 2022). Relegating waste to an Elsewhere entails avoiding responsibility—avoiding paying the costs of the capitalist system's so-called externalities.

The newly emerging awareness goes hand in hand with an increasingly felt need for policies that account for the consequences of processes behind the things we surround ourselves with, their possible harm, and the ways harm is distributed. From their design to every stage of production, distribution and disposal, the objects we consume should therefore be conceived within an ethic of care.

References

Ahmed, Sara. 2017. *Living a feminist life*. Durham: Duke University Press.

Alexander, Catherine, and Patrick O'Hare. 2020. Waste and its disguises: Technologies of (un) knowing. *Ethnos* 88 (3): 419–443.

Arendt, Hannah. 1958. *The human condition*. Chicago: University of Chicago Press.

Armiero, Mario. 2021. *Wasteocene: Stories from the global dump*. Cambridge: Cambridge University Press.
Art Documentary. 2021. *Egypt: Rags and riches*. https://www.youtube.com/watch?v=17-i0LLVFAY. Accessed 27 Feb 2023.
Bailey, Alison. 2021. *The weight of whiteness: A feminist engagement with privilege, race, and ignorance*. Lanham: Lexington Book.
Baraitser, Lisa. 2017. *Enduring time*. London: Bloomsbury Academic.
Bauman, Zygmunt. 2003. *Wasted lives: Modernity and its outcasts*. London: Polity Press.
Calvino, Italo. 2010/1972. *Invisible cities*. Columbus: Vintage digital.
Campbell, Ryan C., and Denise Wilson. 2017. Engineers' responsibilities for global electronic waste: Exploring engineering student writing through a care ethics lens. *Science and Engineering Ethics* 23 (2): 591–622.
Citton, Yves. 2021/2014. Pour une écologie de l'attention. Paris: Points.
Crawford, Kate. 2021. *Atlas of AI: Power, politics, and the planetary costs of artificial intelligence*. New Haven: Yale University Press.
Davies, Thom. 2022. Slow violence and toxic geographies: 'Out of sight' to whom? *Environment and Planning: Politics and Space* 40 (2): 409–427.
Denis, Jérôme, and David Pontille. 2022. *Le soin des choses: Politiques de la maintenance*. Paris: La Découverte.
Douglas, Mary. 1966. *Purity and danger. An analysis of the concepts of pollution and taboo*. London: Routledge & Kegan Paul PLC.
Fineman, Martha. 2005. *The autonomy myth: A theory of dependency*. New York: The New Press.
FitzGerald, Maggie. 2022. *Care and the pluriverse: Rethinking global ethics*. Bristol: Bristol University Press. Kindle.
Franck, Georg. 2005. Mental capitalism. In *What people want: Populism in architecture and design*, ed. Michael Shamiyeh, 98–115. Boston: Birkhäuser.
Fricker, Miranda. 2007. *Epistemic injustice: Power and the ethics of knowing*. Oxford: Clarendon Press.
Gille, Zsuzsa, and Josh Lepawsky. 2021. *The Routledge handbook of waste studies*. New York: Taylor and Francis. Kindle.
Haraway, Donna. 2016. *Staying with the trouble: Making kin in the Chthulucene*. Durham: Duke University Press.
Hawkins, Gay. 2006. *The ethics of waste*. Oxford: Rowman & Littlefield.
Hecht, Gabrielle. 2018. Interscalar vehicles for an African Anthropocene: On waste, temporality and violence. *Cultural Anthropology* 33 (1): 109–141.
Hird, Myra. 2022. *A public sociology of waste*. Bristol: Bristol University Press.
Honig, Bonnie. 2017. *Public things. Democracy in disrepair*. New York: Fordham University Press.
Kaza, S. et al. 2018. *What a waste 2.0: A global snapshot of solid waste management to 2050*. Washington DC: The World Bank. https://doi.org/10.1596/978-1-4648-1329-0
Latouche, Serge. 2012. *Bon pour la casse. Les déraisons de l'obsolescence programmée*. Éditions Paris: Les Liens qui libèrent.
Latour, Bruno. 1993. *We have never been modern*. Boston: Harvard University Press.
Le Guin, Ursula. 2019. *The carrier bag theory of fiction*. Introduced by Donna Haraway. London: Ignota.
Liboiron, Max. 2021. *Pollution is colonialism*. Durham: Duke University Press.
Liboiron, Max, and Josh Lepawsky. 2022. *Discard studies: Wasting, systems, and power*. Cambridge, MA: The MIT Press.
Lynch, Kathleen. 2022. *Care and capitalism*. Cambridge: Polity Press.
Malm, Andreas. 2016. *Fossil capitalism: The rise of steam capitalism and the roots of global warming*. London: Verso.
McIntosh, Peggy. 2020. *On privilege, fraudulence, and teaching as learning. Selected essays 1981–2019*. New York: Routledge.

Medina, José. 2013. *The epistemology of resistance: Gender and racial oppression, epistemic injustice, and resistant imaginations*. Oxford: Oxford University Press.
Molinier, Pascale. 2018. *Le care monde: Trois essais de psychologie sociale; tr. it. Care: prendersi cura. Un lavoro inestimabile*. Bergamo: Moretti & Vitali.
Moore, Jason. 2016. *Anthropocene or Capitalocene? Nature, history and the crisis of capitalism*. Oakland: PM Press.
Morton, Timothy. 2013. *Hyperobjects*. Minneapolis: University of Minnesota Press.
Murphy, Michelle. 2018. Against population, towards alterlife. In *Making kin not population*, ed. Adele Clarke and Donna Haraway, 101–125. Chicago: Prickly Paradigm Press.
Nixon, Rob. 2013. *Slow violence and the environmentalism of the poor*. Cambridge, MA: Harvard University Press.
Oreskes, Naomi, and Eric M. Conway. 2011. *Merchants of doubt: How a handful of scientists obscured the truth on issues from tobacco smoke to global warming*. London: Bloomsbury.
Pease, Bob. 2020. From privileged irresponsibility to shared responsibility for social justice: The contribution of Joan Tronto and iris Marion young to critical pedagogies of privilege. In *The Routledge handbook of critical pedagogies for social work*, ed. Christine Morley, Phillip Ablett, Carolyn Noble, and Stephen Cowden, 165–174. New York: Routledge.
Perkin, Tim. 2017. Enterprising opportunities: Responsibilising Ghana's poor. *Culture and capitalism. A Sussex University anthropology blog*. 27 February 2017. https://cultureandcapitalismblog.wordpress.com/2017/02/27/enterprising-opportunities-responsibilising-ghanas-poor/. Accessed 24 June 2024.
Plumwood, Val. 2001. *Environmental culture: The ecological crisis of reason*. London: Routledge.
Prattes, Riikka. 2020. 'I don't clean up after myself': Epistemic ignorance, responsibility and the politics of the outsourcing of domestic cleaning. *Feminist Theory* 21 (1): 25–45.
Proctor, Robert N., and Londa Schiebinger, eds. 2008. *Agnotology: The making and unmaking of ignorance*. Stanford: Stanford University Press.
Stengers, Isabel. 2013/2009. *Au temps des catastrophes. Résister à la barbarie qui vient*. Paris: La Découverte.
Sullivan, Shannon, and Nancy Tuana. 2007. *Race and epistemologies of ignorance*. New York: State University of New York Press.
Tronto, J. 1990. *Chilly racists*. Paper presented to the annual meeting of the American Political Science Association, San Francisco, California, August 30–September 2.
———. 1993. *Moral boundaries: A political argument for an ethics of care*. New York: Routledge.
———. 2013. *Caring democracy: Markets, equality, and justice*. New York: New York University Press.
Veblen, Thornstein. 1899. *The theory of the leisure class*. Oxford: Oxford University Press.
Vergès, Francoise. 2019. Capitalocene, waste, race, and gender. *E-flux Journal* 100 (5/19): 1–13.
Vinsel, Lee, and Andrew L. Russell. 2020. *The innovation delusion: How our obsession with the new has disrupted the work that matters most*. New York: Currency.
Zembylas, Michalinos, Vivienne Bozalek, and Tom Shefer. 2014. Tronto's notion of privileged irresponsibility and the reconceptualisation of care: Implications for critical pedagogies of emotion in higher education. *Gender and Education* 26 (3): 200–214.

Chapter 10
Public Policy on Child Care in Mexico: The Case of Childcare Centers to Support Working Mothers

Urania Lanestosa

Abstract In Mexico, childcare has been recognized as a public problem. The growing incorporation of women into the labor market, along with poverty and inequalities, has made it necessary to design and implement public actions aimed at meeting the childcare demands that women, in particular, face. One of these public actions is the Programa Estancias Infantiles (PEI), which sought to contribute to the generation of female employment and the satisfaction of the care needs of children under 6 years of age. After 12 years of operation, childcare centers in Mexico have faced significant changes. With the cancellation of this program in 2019, a significant number of women who had been dedicated to childcare for years faced reductions in their salaries and, in many cases, lost their jobs due to changes in public policy and the impact of the COVID-19 pandemic. Although the federal government led by Andrés Manuel López Obrador emphasized that these changes were not intended to close childcare centers, existing evidence shows that they did contribute to a decrease in the supply of care services for working mothers who do not have social security and live in conditions of poverty. This has also impacted the reduction of spaces where women can be employed and have a source of income to meet their daily needs. Government decisions at the beginning of the program and its cancellation were not oriented to de-feminize childcare. This text is based on the doctoral research "De beneficiarias a actoras política: análisis feminista de las experiencias de siete responsables de estancias infantiles y de las transformaciones de la política de cuidado infantil en México," which is being developed with the support of the Consejo Nacional de Humanidades, Ciencias y Tecnologías (CONAHCYT).

Keywords Childcare centers · Care work · Feminization of care · Policy

---

U. Lanestosa (✉)
Division of Social Sciences and Humanities, Universidad Autónoma Metropolitana-Xochimilco, Mexico City, Mexico
e-mail: urania1b@gmail.com

J. Braga (ed.), *Ethics of Care*, Advancing Global Bioethics 20,
https://doi.org/10.1007/978-3-031-73608-7_10

1 Child Care and its Connections to Public Policy

The physical, cognitive, social, affective, and emotional well-being of children requires the development of many daily activities, relationships, and bonds. According to Valeria Esquivel (Esquivel et al. 2012), a broad definition of child care allows us to recognize a very diverse repertoire of actions and daily care: feeding infants and young children who cannot feed themselves, bathing them, putting them to bed, reading to them, playing, talking to them, giving them emotional support, coordinating social activities, being aware of what they do, supervising their games, among other activities.

Thus, child care implies allocating time, effort, and attention to ensure that from our first days of life, human beings have the necessary conditions to satisfy vital needs such as food, safety, and affection, among others. Like other forms of care, child care represents work that needs to be valued, appreciated, and recognized. This way of understanding child care allows us to grasp its social, economic, and human value. According to Pascal Molinier:

> Caring for others is not only thinking about them, worrying about them intellectually or even affectively; it is not even necessarily loving them. It is to do something, to produce work that contributes directly to the preservation of the other's life. It is to help or assist them in their primary needs such as eating, being clean, resting, sleeping, feeling safe, and being able to devote themselves to their interests [my translation] (Molinier 2018, 189).

Child care is work that can be paid or unpaid, depending on the social relations that are built between different actors. Thus, it can be performed within families or personal networks without payment; it can also be performed by caregivers who do not have a personal, family, or affective bond and who receive remuneration in exchange for their work or where there is another type of exchange for the work performed.

Most studies on child care have focused on the dynamics, processes, and strategies that occur within the home. However, we should not lose sight of the fact that child care can be carried out in different contexts and spaces. The development of research that allows us to understand how it is carried out in public spaces is of relevant importance for the field of care, particularly in services and centers designed to meet children's daily needs and provide support to families, as well as in community care networks.

According to the study conducted by the National Council for the Evaluation of Social Development Policy, childcare services in Latin America and the Caribbean were conceived as a response to the increased participation of women in the labor market and the assumption that mothers with young children from low economic strata could increase their chances of accessing employment or education (CONEVAL 2022). According to Araujo and López-Boo:

> "the poorest women are those who work the most hours and, also, the women who work the most (or have worked) are those who most demand care services" (Araujo and López-Boo 2015:251) for their daughters and sons, which may also be related to the lack of networks in the households or to the precariousness of employment. However, one of the most

> important findings of this study is that no families were found with access to full- or part-time care services.... [my translation] (CONEVAL 2022, 55).

Child care, in particular, has received significant public attention from different actors (governments, civil society organizations, international agencies, community networks, families, among others), which has favored progress in its recognition as a priority issue. It has also allowed the allocation of public resources so that families have the economic resources to cover the costs involved in its satisfaction, mainly in contexts of poverty, as well as care services in which children can develop comprehensively in safe spaces.

A significant advance is the recognition that care in general, and child care in particular, is a matter that is not only the responsibility of families; it requires the design of care policies understood as:

> public actions aimed at organizing the work of caring for and assisting people in situations of dependency—so that they can carry out the basic, instrumental, and advanced activities of daily living—with the caregivers also being the subjects of these policies. This concept includes measures that guarantee access to services, time, and resources so that people can care for and be cared for in conditions of quality and equality [my translation] (ONU Mujeres-CEPAL 2020, 2).

Despite these advances, in the Latin American region, there is a strong maternalist and familist culture present not only in governmental actors but also in society. Following Eleonor Faur (in Esquivel et al. 2012), maternalism is a perspective that assumes women as "mothers" and as "the best caregivers of their children."

> Caregiving has historically been an activity assigned to families, but more precisely to mothers, who were and still are the main caregivers in the home. This vision was accompanied by a social policy model, an institutional organization, and a system of rights that had as an assumption a type of family configuration with a male income provider and a female full-time homemaker who cared for children, the elderly, and the sick [my translation] (Faur 2015, 2).

The best child care continues to be associated with that provided in the home by the women in the family (mothers, grandmothers, aunts), which adds to the centrality that the notion of the traditional nuclear "family" still has in the collective imagination, even though in practice the diversity of family arrangements is increasingly visible in everyday life.

Maternalism and familism present significant challenges for the design and implementation of public policies. They highlight the need for society to strongly advocate for the expansion of public childcare services by the state and to promote greater involvement of diverse social actors in their provision. In public policy, it is essential to incorporate a critical perspective on care to foster changes in how daily needs are addressed and in the roles of those responsible for caregiving.

2 Child Care in Mexico: The Case of the Estancias Infantiles Program (PEI)

In Mexico, caregiving is still not a right recognized in our national Constitution. Although since 2020, the bill to reform Articles 4 and 73 of the Constitution was approved by the Chamber of Deputies for review by the Senate, which would allow for a National Care System (Senate of the Republic 2021), this initiative has not made significant progress to date.

Both the public and private sectors provide childcare services. The provision of public childcare, care, and comprehensive child development services depends on the type of social security scheme in which the mother or father is enrolled (CONEVAL 2019). Access to public childcare services is largely conditioned by the existence of a formal job, social security, or sufficient economic resources to pay for private services. This has historically disadvantaged many women who have children between 1 and 6 years of age, and who do not have a family or community network and/or an income that allows them to meet their care and/or social security needs.

According to the General Law for the Provision of Services for the Attention, Care, and Integral Development of Children (LGPSACDII) in Mexico, Childcare Centers (Centros de Atención Infantil, CAI) are public, private, or mixed modality spaces where services are provided for the attention, care, and integral development of children from 43 days old to 6 years old. These services can be of three types: public, subrogated, and private services; community care; and home visits (Sistema Nacional para el Desarrollo Integral de la Familia 2020).

In practice, care networks are mostly supported by families or by great personal efforts to meet care needs in contexts of deep inequalities. Public childcare services are largely determined by the existence of formal employment and social security, leaving out many families and women who, although they want to study or work outside the home, cannot do so because of the time and significant work they devote daily to caring for their children, as well as other people in their homes who require attention and support to meet their daily needs.

Childcare has seen significant development and increased public attention through the establishment of public programs and initiatives specifically designed to support its provision. One such program is the Programa de Estancias Infantiles (PEI), which was introduced in 2007 during Felipe Calderón's administration. This program aimed to increase female employment and address the childcare needs of children aged 1 to 4. It sought to boost the participation of 'working mothers' and 'single-parent families' in the labor market by offering childcare services and expanding access to these services (SEDESOL 2007).

The PEI allowed women in conditions of poverty and marginalization to have access to care services. It also motivated women who were unemployed or who were looking for opportunities for paid work to find child care as an activity that they could carry out at home. The then president of Mexico, Felipe Calderón, expressed in different governmental messages that during his campaign around the country, he became aware of the needs of two groups of women: "single mothers" who could not go out to work because they had no one to leave their children with, and women who also wanted to have a job but did not have opportunities to do so and faced discrimination due to their age.

Women from all over the country found in this government initiative opportunities to become responsible for childcare centers and support other women in communities where there were significant childcare needs and poverty. The notion of the woman-caregiver was present from the origins of this program, reproducing maternalism in the way child care is often viewed. Although the initial rules of operation refer to the fact that it could not necessarily be women who had this role, the governmental narratives of then President Felipe Calderón exalted at different times the participation of women in child care and the installation of child care centers in their own homes.

> We are going to give her (the woman) a loan to fix up her house, to install a bath for the children, to protect them from entering the kitchen, to separate the childcare area from the rest of the family's activities, and to build benches, chairs, and a children's room [my translation] (La Jornada 2007a).
>
> This program is conceived with a different logic and common sense. We are not thinking about large facilities, big rooms, or grand architects; no, we are thinking about houses, patios, garages, living rooms, and dining rooms, which can be adapted to be excellent childcare centers—for whom? For the children of the neighborhood, whose mothers ahora están trabajando, a pesar de que no era socialmente habitual que trabajaran en el pasado [my translation] (La Jornada 2007b).

During the time this program was in operation (2007–2019), it benefited 1,825,394 single mothers and fathers, provided care to 2,174,415 children, and had 9566 childcare centers in operation with national coverage. The implementation of the PEI was characterized by the provision of public resources to establish and operate a childcare center affiliated with a national network.

Each childcare center had to have a person in charge and, depending on demand could hire personnel, primarily educational assistants, to provide care services. The Federal Government, through SEDESOL, made payments directly to the childcare center (initially 700 pesos per month was granted for each child from 1 to 6 years of age, which is currently equivalent to US$37.85). Additionally, the mother or father was required to make a monthly co-responsibility contribution equivalent to 10% of the amount of federal support per child enrolled, which was paid directly to the center.

At the start of the PEI, individuals living in poverty who wanted to establish and operate a childcare center could receive up to 35,000 pesos (currently equivalent to US$1892) to adapt and equip the facility, train administrative and operational staff, and purchase or prepare materials for working with the children. They could allocate up to 20% of these funds to cover operating expenses for the first 2 months (SEDESOL 2007).

A relevant question in the context of this program is why install childcare centers in family homes or properties? It was an inexpensive, "quick solution" that did not require significant effort from the government. This approach considered that there were mothers who owned homes where their children were no longer living, so if they received resources for adaptations, they could create childcare spaces and simultaneously have a source of employment.

Another reason related to the government's choice was the belief that homes are ideal spaces for childcare. Although there was recognition of the need to consider safety aspects in the installation of childcare centers, at the program's inception, it was thought that simple and quick adaptations would be sufficient.

As a result, many women set up childcare centers in their own homes or family properties, blurring the boundaries between public and private. Where does the domestic sphere end, and where does the workspace begin? How does paid childcare coexist in the same time, space, and corporeality as the unpaid care that these women perform daily in their homes?

During fieldwork conducted with seven centers across different contexts and locations in Mexico, I found that only two centers were established in rented spaces. The remaining centers were run from the managers' own properties, leading to especially long working hours. In the remaining five cases, the childcare center managers negotiated with their families and made significant material changes to adapt their homes for the care of children from their communities. This included converting rooms, dining areas, bathrooms, and previously shared spaces into dedicated childcare facilities.

These transformations brought about significant changes in family dynamics, work schedules, rest, and personal time. For women who established childcare centers in their homes or family properties, working hours were particularly long. They were responsible for labor demands from early morning to late evening, which included food preparation, purchasing supplies, and maintaining and improving the centers.

Over time, childcare center managers had to meet increasingly stringent standards to remain in the program. Specifically, aspects such as space, staffing, capacity, and safety conditions were scrutinized more closely. The limited income they earned was used to support themselves and their families, as well as to upgrade facilities to better meet government requirements and increase the center's capacity.

Although the seven women who participated in the research have different life trajectories and come from diverse backgrounds—three from various regions of Mexico City, one from the State of Mexico, one from the State of Sinaloa, one from the highlands of Querétaro, and one from the State of Hidalgo—they share several common traits. All are mothers aged between 56 and 62 years.

They expressed a deep passion for education and working with children, as well as a strong social commitment to helping others in their communities, particularly women and children. They come from diverse backgrounds and most hold a bachelor's degree. Throughout their participation in the program, they took on significant childcare responsibilities and needed work that they enjoyed and that allowed them to stay connected to their homes.

An important finding from the research is the development of strategies that extended beyond the program's rules. In all cases, childcare center managers established community care networks that supported the local solidarity economy (purchasing food and groceries from small local stores and receiving food donations). One caregiver's testimony underscores the significance of these cooperative care relationships.

> In our community, we also started to create solidarity networks. What do I mean by solidarity networks? Instead of buying from the supermarket, I would buy from the local grocer, or the tortilla shop, or the butcher. This way, we built solidarity networks and connections within our community. Often, people would say, "Teacher, I have some leftover vegetables, would you like some for the children?" or "Teacher, I have some leftover chicken..." Well, "Give it to me." It wasn't always the case, but we did receive support from the same community that was involved in caring for those children [my translation] (Responsible for the childcare center, Mexico City).

Another aspect to highlight is that the women childcare center managers became important references for other women who came to them for advice on what to do if they were experiencing violence in their homes or facing other problems they did not know how to resolve. These cooperative relationships were not contemplated by the program but were something that the women developed on their own.

> We saw that we could address more issues, not just as a childcare center, because you become a prominent figure in your area. People then think that you have great influence, but I told them that I receive supervision [...] It's great because women come to you for support [my translation] (Responsible for the childcare center, Mexico City).

Although the PEI did not set out to profoundly transform how society assigns the primary responsibility for child care to women, it did bring to the public stage the fact that women in poverty, who lack jobs and social security, require additional public support to cover the costs of child care, as well as access to a network of child care services. Thus, the PEI recognized the need for two important elements of childcare: financial support and child care services in impoverished communities.

3 Conditions for Care and Self-Care

Reviewing the conditions under which people perform their work is crucial for understanding how it is carried out and the value society places on it. This is especially relevant for care work, as, despite its importance for sustaining life, it has historically been regarded as a female activity and undervalued.

This devaluation of care work is not only present within the home but also in the public sphere. This leads women, such as those managing childcare centers, to face challenges in securing a living wage, adequate working conditions, and social recognition. According to Judith Butler:

> People who care for others not only take care of other people; they also need to have their own support needs met (i.e., decent conditions in terms of work and rest, wages, housing, and medical care). The conditions of support for the most vulnerable moments of life are themselves vulnerable, partly due to infrastructural issues and partly due to human and technical elements [my translation] (Butler 2017, 132–133).

When I started the research, I initially thought that these were women engaged in paid child care. However, as I spoke with the women managing the seven childcare centers included in the research, my perception gradually changed. For these women, remuneration for their work was more of a hope than a reality.

The operation of the PEI was based on a model where those in charge were expected to generate their income without receiving direct remuneration or payment from public resources. They were also not included in the social security system. Consequently, payment for their work was not guaranteed and depended on the internal organization they established at each center, the resources available, and the expenses they had to cover for operations.

> Do you know how much salary we pay them? Nothing, no salary. The person in charge of the center has to make arrangements so that the mothers accept and prefer the one we support. The mother receives a card with a monthly amount and uses it to pay for the childcare service. If the mother is not satisfied with the center or if the care is inadequate, she can choose another one. Therefore, the childcare center managers must provide good care, keep the children happy, and do their job well. This is another key to the success of the program. Even though we do not pay any salary to the center managers, and hence there is no labor relationship or union [...] we have no payroll, which allows us to expand the centers... [my translation] (Presidency Felipe Calderón Hinojosa 2012).

As a result, the responsibility for compensating their work shifted to these women, depending on the quality of the care service they provided and their ability to attract and retain families. These savings in labor costs enabled care centers to expand across the country, despite limited public spending.

What motivated these women to work in child care without a guaranteed salary? Initially, the government's narrative had a significant impact not only on the women themselves but also on those close to them, who saw the PEI as a promising opportunity. It was an activity they could undertake from home, with some support from the government.

Another reason is related to the women's own economic and care needs. Three of the seven women in charge explicitly stated that they were looking for a job that they liked and that would allow them to be closer to their children. What they all agreed on was that they were interested in helping their communities, mainly young children and other women.

> We thought it was to help, didn't we? For the good of the community, of course when we saw that it was our income and that it was our job and that we generated it and that we had this income that was self-employment, well, the truth is that I did not get rich from this, it was my self-employment that generated jobs for other women, you take care of it and you take responsibility, until they took us out and it was terrible but if you think you have to help other women, you think you have to help them. I think that if you think you have to help other women: from the grandmother who also had her grandson thrown out, who doesn't have a birth certificate, you go around helping everyone because you believe it and it feels great [my translation] (Responsible for the childcare center, Mexico City).

To raise sufficient resources and guarantee the operation of the daycare centers, those responsible had to make significant efforts in their communities to publicize their services so that, gradually, more families would bring their children. In seeking this approach in their communities, they observed that there were significant child care needs in addition to contexts of social deprivation, violence, and insecurity.

> Many children were abused at home, they were beaten, they were brought there, I don't know, I saw that there were mothers who sold, they were street vendors, I saw a mother with a tricycle, she sold shaved ice and, like this on the ice, she put her ice bar and it dripped to the bottom and on the bottom she brought the baby lying there: very dirty, wet.... [my translation] (Responsible for the childcare center, State of Mexico).

When they saw the conditions of poverty in their community contexts, they generated strategies so that families with fewer resources would not pay any additional fees, even if this decision implied a reduction in their income. In contrast, in the cases of families with greater economic resources, they raised the additional fee they had to cover to compensate those who could not make this extra payment and whose care was covered only with government support.

The interesting thing about these strategies is that they were not contemplated in the program's operating rules; it was a decision made by those responsible and was common in many of the childcare centers.

> Many of us teachers had many "scholarship" children, because we saw that the mother was a street sweeper, a domestic worker, guía de estacionamiento informal, selling on the street, because we saw that the mother was suffering, so we said "how am I going to charge her?", but suddenly someone would arrive with a big car and say "I want my child to be here in the daycare" and with very good economic possibilities "because I know you, teacher, I know your work and because I am going to see if they give me the support", "well go and apply for it". And, "Well, that's up to them", and if they did a socioeconomic study and accepted her, I don't care, the child is already there. So I would go up a little bit, to that lady, I would ask her for a quota that was almost equal to the SEDESOL quota, and this quota was useful for the one that didn't give me anything [my translation] (Responsible for the childcare center, Mexico City).

In addition, these women had to manage the public resources they received according to the number of children they cared for and the maximum capacity that, according to the physical conditions of the space, the authorities allowed them to receive. With this money, they had to pay the salaries of those who worked with them, pay for electricity and water, maintain the childcare center in optimal conditions, and comply with the requirements of the periodic supervision carried out by the authorities. The salary for their work could not always be paid.

> Yes, I had a salary, but my salary was the last one that was paid. Everything was paid first. Because there were times when there were needs at the children's center. Every year, when you had to renew your internal plans, that is, you had to invest because nobody supported you, nobody came to give you anything, from painting, every year you had to waterproof, to maintain the water tanks, the electrical inspections, the reports you needed to be able to renew all your plans. You had to spend a lot on that. So I knew that there were times of the year when my salary was not going to be enough because I had to cover all of that [my translation] (Responsible for the childcare center, Sinaloa).

Another important condition for valuing the work of the women in charge has to do with their working hours. Most of them had a long working day since they were open for an average of 8 h. The operation of the centers took into account the needs of each context and the times when families required support to carry out their work activities, as in the case of some childcare centers located on the northern border of Mexico, which had a night schedule and began to provide services at 8:00 pm.

Additionally, the women in charge of the childcare center felt immense pressure to be on-site at all times to ensure everything was in order, as authorities could arrive at any moment to supervise operations.

> I never went out and left the childcare center. You know what I was afraid of? Not only did we not have permission to leave, but if we were going to leave, we had to notify in advance, as if illnesses would warn us: "By December 14, you are going to be sick." They treated us like that and would not let us go out. If there was a call and we were in the bathroom, they would say, "Oh, she went out, she went out because she did not answer the phone." It was tremendous harassment at work, tremendous... [my translation] (Responsible for the childcare center, Mexico City).

Once the daycare closed, they continued working: preparing materials for the next day, going out to buy food to prepare meals for the children, reviewing teaching plans, and ensuring everything was ready to start activities day after day. In some cases, the stress, workload, lack of rest, and insufficient conditions for self-care led some of these women to fall ill and require emergency care, despite not having medical coverage. Without access to social security, they were unable to access public healthcare.

> It was a lot of stress, this insomnia too, because of worries, things that had to be accomplished, and thinking about how am I going to solve them? My economy was much better before starting the center, I have to tell you. For a while, I kept going because it was the product of my work, and when I realized that it was no longer enough, I went back to developing productive projects, I became active as a thesis advisor and went back to giving workshops [my translation] (Responsible for the childcare center, Sinaloa).

As can be seen, participating as responsible for childcare centers was a demanding job that consumed many hours of their working day and to which the seven women who participated in the research dedicated between 5 and 12 years of their lives. Their work gave them great satisfaction, but it was not free of tensions, difficulties, and affectations.

4 How Do Changes in Public Childcare Policy Affect Women?

After 12 years of operation, childcare centers in Mexico faced significant changes that posed challenges to their sustainability. In 2019, the government under President Andrés Manuel López Obrador announced a major modification shortly after coming to power: the direct delivery of public resources to mothers, single parents, or guardians, allowing them to decide how to manage these funds. Consequently, childcare centers stopped receiving public resources for their creation, operation, improvement, and staff training, as well as other support essential for their operation.

On February 28, 2019, the Official Gazette of the Federation published the rules of operation for the Programa de Apoyo para el Bienestar de las Niñas y Niños, Hijos de Madres Trabajadoras (PABNNMT), which replaced the PEI (Secretaría de Bienestar 2019). In this new program, those responsible for childcare centers were not considered relevant actors in meeting care needs. The rules did not explicitly consider these care centers as services requiring co-responsibility and cooperation between the government, the caregivers, and the families.

In various cities across the country, women organized public demonstrations with two main goals: to highlight their work, which had remained hidden behind the doors of the childcare centers for over a decade, and to demand dialogue with the government, particularly with the president, who showed no interest in engaging. Their call to the government went unanswered.

Although the federal government emphasized that these changes were not intended to close daycare centers, existing evidence shows that they did contribute to a decrease in care services. This reduction affected both working mothers without social security and other families using these services. Additionally, these changes diminished employment opportunities for women who had been involved in childcare in their communities for over a decade. This situation has impacted their ability to earn income and meet their daily needs.

The government's justification for these changes focused on common elements in public management of social programs: reducing public spending, directly delivering resources to beneficiaries to avoid corruption, and rejecting actions and interventions deemed neoliberal by the current administration, which had been implemented in previous administrations. Additionally, the government cited inconsistencies (such as children not attending daycare centers, non-functional centers, and unregistered address changes) and alleged irregularities in the management of public resources. However, the government has yet to provide evidence showing which centers were found to have irregularities.

Of the seven childcare centers included in the study, four have closed permanently. Two now cater exclusively to families who can afford to pay privately for their services. The remaining center continues to operate with the support of its local community government. Those in charge have had to rely on their savings and family resources to sustain themselves. The families who attend these centers have also changed; they now serve families who need services but do not receive government assistance to cover the costs.

The caregivers involved in the investigation have shared feelings of betrayal. They trusted the current president and supported him during his campaign, expecting positive changes. Instead, they now face more complex conditions than before.

> It destroyed us, it annihilated us, and it annihilated us for no reason because we simply needed to be audited, childcare center by childcare center, and whoever needed to be punished should be punished [my translation] (Responsible for the childcare center, Mexico City).

The poverty and precarious conditions of the families meant that, instead of continuing to take their children to childcare centers, they used public resources to satisfy other needs. In the seven centers that are part of the investigation, those in charge reached agreements with the families to continue providing the service even though the families could not make any payments, with the promise that when they received governmental support, they would settle the outstanding debt. However, this did not happen.

Those responsible for childcare centers had to bear the financial costs of regular operation and even incurred debts they could not repay. In all cases, they had to reduce the number of staff, as they had no money to pay salaries.

> Imagine a father in dire need who receives 3200 pesos in his hand. It is a lot of money for him, a lot. Then they leave, and we are left with this heavy debt: we owe the rent, we often owe for groceries, electricity, so we don't have enough... half of our population leaves, and we don't have enough resources to continue covering our needs [my translation] (Responsible for the child care center, Mexico City).

In Mexico, as in other parts of the world, the COVID-19 pandemic brought increased attention to the importance of care; discussing care became more common and significant. Childcare centers that were still trying to survive had to close due to the increase in contagions and because they were not considered essential services. Paradoxically, the concerns of citizens and public opinion were more focused on the evolution of the pandemic, the number of infections and deaths, prevention measures, and the availability of hospitals and centers for illness, among other aspects. No coverage was given to the problems that the closure of childcare services represented for mothers, nor was it possible to understand how those in charge of childcare centers were managing during this difficult period.

Among the many questions we can ask about the impacts of governmental decisions, two seem central: what effects do these public policy decisions have on women's lives, and to what extent do these transformations represent a change in how care is viewed by governmental actors and society?

While the direct provision of resources represents a significant change, the primary responsibility for care, which falls on women, has remained unchanged. Despite the closure of numerous daycare centers, the national government has not implemented specific measures aimed at supporting out-of-home care services. By exclusively focusing on familial care, there is a risk of perpetuating inequalities and power dynamics that disproportionately affect women, and of reinforcing the notion that the best care is provided by mothers and grandmothers within the household.

In addition to the economic impact, many women have experienced a long period of desperation, searching for alternatives to continue their work and recover from the emotional blow caused by these changes and accusations of corruption. According to those responsible who have participated in the investigation, those who have not been able to reopen their centers are now selling goods or have adapted their care centers into spaces where they prepare food to survive.

5 Final Thoughts

Analyzing programs such as the PEI provides evidence that public and governmental decisions can generate multiple changes in women's lives and the management of care, some of them positive and some negative. These analyses also reveal strategies of resistance and women's capacity for agency to turn the care we have historically provided within the home into relevant political issues.

Studying programs that have been canceled or undergone significant changes might seem like a fruitless task. However, in my research experience, I have recognized that such analyses can contribute to public decisions that consider the implications for the lives of the people involved. The conclusion of a program responds to administrative, contextual, and political logic, but many of the practices, meanings, and experiences prevail much longer.

Public interventions can address various aspects, with childcare services being a crucial focus. These services provide evidence of the significance of having material, human, and financial resources that enable women to actively participate in the labor market and broaden their educational opportunities.

Social support for child care is an important part of addressing some of the needs that families face, particularly those associated with the costs of care. However, care decisions in households where there is poverty, precariousness, violence, and multiple limitations have particularities that need to be considered by public decisions. In many cases, these decisions can perpetuate inequalities in the distribution of care responsibilities among household members.

In countries such as Mexico, where access to care services is linked to having a formal job and social security, many children see their opportunities to grow up in spaces where their safety, development, and care are a public priority reduced. For many women, caring for their children, grandchildren, nieces, and nephews in their homes is not a decision made reflexively; it implies fulfilling a social mandate that sometimes distances them from their desires and needs.

In Mexico, the cancellation of support for childcare centers and the social, economic, and health repercussions of the COVID-19 pandemic have created a complex context for childcare that requires more knowledge and research. This experience highlights the need for care policies to include actions that recognize the important work of those responsible for childcare centers in their communities. Additionally, the design and budgeting of such programs should consider fair remuneration and the necessary conditions not only for child welfare but also for the development of care work, considering caregivers' needs and access to other rights such as social security, medical care, rest time, and the necessary support for their professionalization.

Care and the value of the work performed by caregivers is a political struggle that does not follow a linear and progressive path, but one with slopes and obstacles. Generating changes in the way childcare needs are met in societies such as Mexico's is a long process requiring the participation of different actors to expand existing services. It also requires recognizing that care is not solely a family matter but also a responsibility of the State, the market, and communities.

Care research not only broadens our knowledge and understanding of what care is, what activities it involves, who participates, and the relationships that are built, but also provides a space to name, recognize, and keep alive in memory the work and political action of women, such as those responsible for childcare centers, who make daily contributions to their communities, often sacrificing their self-care and well-being. Is also a space to generate networks and links, making care a collective struggle in which the experiences of those who develop it inside and outside the home build valuable, diverse, and necessary knowledge and experiences to sustain life.

References

Araujo, María Caridad and Florencia López-Boo. 2015. Los servicios de cuidado infantil en América Latina y El Caribe, in *El trimestre económico*, (2)326, pp. 249–275. Ciudad de México.

Butler, Judith. 2017. Vulnerabilidad del cuerpo y la política de coaliciones. In *Cuerpos aliados y lucha política. Hacia una teoría performativa de la asamblea*. Trans. Maria José Viejo, 125–156. Barcelona: Paidós.

CONEVAL. 2019. *Programa de Estancias Infantiles Para Apoyar a Madres Trabajadoras. Recuento de La Evidencia de Las Evaluaciones 2007–2019*. https://www.coneval.org.mx/Evaluacion/RDPS/Documents/NOTA_ESTANCIAS_INFANTILES.pdf. Accessed 28 Feb 2023.

———. 2022. *Pobreza y cuidado infantil: Un estudio cualitativo en hogares en México*. https://www.coneval.org.mx/Evaluacion/IEPSM/Documents/Informe_pobreza_y_cuidado_infantil_021222_.pdf. Accessed 28 Feb 2023.

Esquivel, Valeria, Eleonor Faur and Elizabeth Jelin. 2012. *Las lógicas del cuidado infantil. Entre las familias, el estado y el mercado*. Argentina: Ides-UNFPA-UNICEF.

Faur, Eleonor. 2015. *La organización social del cuidado infantil*. Diálogos del SIPI. Conversación con Eleanor Faur. http://equidadparalainfancia.org/wp-content/uploads/2016/03/SIPI-2015-Conversaci%C3%B3n-con-EleonorFaur.pdf. Accessed 28 Feb 2023.

La Jornada. 2007a. Comienzan a operar las estancias infantiles anunciadas. *Claudia, Herrera Andrés T., Morales*. www.jornada.com.mx/2007/01/12/index.php?sección=sociedad&artículo=043n2soc. Accessed 28 Feb 2023.

———. 2007b. *Defiende Felipe Calderón su programa de estancias infantiles*. https://www.jornada.com.mx/2007/07/26/index.php?sección=sociedad&artículo=035n2soc. Accessed 28 Feb 2023.

Molinier, Pascale. 2018. El cuidado puesto a prueba por el trabajo vulnerabilidades cruzadas y saber-hacer discreto. In *El trabajo de cuidados*, 187–210. Buenos Aires: Fundación Medifé Edita.

ONU Mujeres-CEPAL. 2020. Cuidados en América Latina y el Caribe en tiempos de COVID-19. Hacia sistemas integrales para fortalecer la respuesta y la recuperación. https://www.cepal.org/es/publicaciones/45916-cuidados-america-latina-caribe-tiempos-covid-19-sistemas-integrales-fortalecer. Accessed 28 Feb 2023.

Presidencia Felipe Calderón Hinojosa. 2012. Quinto Aniversario de Estancias Infantiles Felipe Carrillo Puerto, Quintana Roo. https://www.youtube.com/watch? v=mOpasfejLJo. Accessed 28 Feb 2023.

Secretaría de Bienestar. 2019. *Acuerdo por el que se emiten las reglas de operación del programa de apoyo para el bienestar de las niñas y niños, hijos de madres trabajadoras, para el ejercicio fiscal 2020.* Diario Oficial de La Federación. https://www.dof.gob.mx/nota_detalle.php? Código=5583303&fecha=31/12/2019. Accessed 28 Feb 2023.

SEDESOL. 2007. *Acuerdo por el que se emiten y publican las reglas de operación del programa de guarderías y estancias infantiles para apoyar a madres trabajadoras, para el ejercicio fiscal 2007.* Diario Oficial de La Federación. https://www.dof.gob.mx/nota_detalle.php? Código=5012023&fecha=30/12/2007#gsc.tab=0. Accessed 28 Feb 2023.

Senado de la República. 2021. *Sistema nacional de cuidados, fundamental para la igualdad de derechos, afirman en el Senado.* Diario Oficial de La Federación. http://comunicacion.senado.gob.mx/index.php/informacion/boletines/50745-sistema-nacional-de-cuidados-fundamental-para-la-igualdad-de-derechos-afirman-en-el-senado.html. Accessed 28 Feb 2023.

Sistema Nacional para el Desarrollo Integral de la Familia. 2020. *Programa Nacional de Prestación de Servicios para la atención, cuidado y desarrollo integral infantil (PNPS) 2020–2024.* Diario Oficial de La Federación. Ciudad de México. https://www.dof.gob.mx/nota_detalle.php? Código=5608937&fecha=28/12/2020&print=verdadero. Accessed 28 Feb 2023.

Chapter 11
Informal Care: Public Care Policies Versus the Privatisation of Family Labour

Marcela Uchôa and Jonas Van Vossole

Abstract The March 2020 implementation of the Informal Caregiver Statute in Portugal at the onset of the pandemic revives the need to debate the condition of women in caregiver roles, as well as the role of care and domestic work in capitalist societies. Thinking through the feminist lens on primitive accumulation that affects semi-peripheral societies like Portugal helps us to understand ambiguous positions regarding public policies. These policies are still proposed as guarantees for women to perform jobs with minimum security, jobs that actually allow for privatising a responsibility of the State. In this article, we mean to contextualise and problematise the Informal Caregiver Statute through a discussion about capitalism, patriarchy, the sexual division of labour and the demise of the welfare state.

Keywords Care · Social reproduction · Capitalism · Women

1 Introduction

In this article, the recent legislative changes regarding the new Informal Caregiver Statute in Portugal serve as a reference point for a much-needed discussion about politics of care, capitalism, welfare and the sexual division of labour. While at first sight the implementation of this statute seems to be progressive (providing vital financial aid, legal protection and recognition to informal caregivers) with some well-known feminists having advocated for similar measures, we will argue that this legal change is problematic from a feminist perspective.

M. Uchôa (✉)
Faculty of Arts and Humanities, Institute for Philosophical Studies, Coimbra University, Coimbra, Portugal

J. Van Vossole
Ecology and Society Lab, Center for Social Studies, Coimbra University, Coimbra, Portugal
e-mail: jonasvanvossole@ces.uc.pt

J. Braga (ed.), *Ethics of Care*, Advancing Global Bioethics 20,
https://doi.org/10.1007/978-3-031-73608-7_11

In the first part of this article, we will analyse the context and the legal details of the proposed new statute, as well as some of the problems. In the second part, we will contextualise the legal change within the political, economic and historical context of care. We will particularly focus on the relation between care, patriarchy and the sexual division of labour, which has tended to delegate the unpaid labour of social reproduction to women. We will then analyse the relation of care in the context of the social welfare state and its demise. Finally, we will see how social and economic crises under neoliberalism and the pandemic affect care, as well as potential alternatives for care in crisis.

2 A New Statute of Informal Care

Before the pandemic, in Portugal, approximately 240,000 people lived in a situation of dependence. Like in most places with ageing populations, these numbers have generally increased as people tend to live longer due to scientific advances. If such extension of life is to be accompanied by true concerns about maintaining or improving the quality of life, there must be changes to the existing mechanisms and institutions of care. In this context, profound reflection is needed on the policies accompanying this evolution and their effects, as well as on the logics of social reproduction under capitalism, in general.

This need increased in urgency due to the onset of the pandemic crisis, which occurred around the same time as the official publication of the care statute. In a few months, the number of citizens who provide care for dependent people doubled. With less than 10 million inhabitants in Portugal, the number of informal caregivers was actually expected to be around 1.4 million, according to a survey carried out by the National Association of Informal Caregivers. The number surged during the pandemic, due to the lack of social responses, closure or isolation of existing care institutions, closure of schools, nurseries, restaurants, etc.

Just before the onset of the pandemic, policymakers tried to address this problem through the legislation of the so-called Informal Caregiver Statute. Initially, the legislative proposal would create a legal framework for those people who spent their days caring for others. It had to address economic difficulties for those caregivers by providing a financial subsidy, as well as guaranteeing some social rights. The measure had been supported by the parliamentary left majority, composed of a PS-minority with support from the Communist Party and the Left Bloc. In particular, the Left Bloc Member of the European Parliament, Marisa Matias, had been a vocal proponent of a similar measure for some time. Also, right-wing parties, such as CDS-PP and PSD, gave several contributions to the new statute from a Christian caritative perspective.

2.1 The New Law

In September 2019, this resulted in a new Informal Caregiver Statute in Portugal, which was legislated and published as an annex to Law No. 100/2019. The ordinance defined the terms and conditions for implementing pilot projects provided for in the Informal Caregiver Statute. Despite having officially entered into force on September 7, this ordinance was only officially published in *Diário da República* on March 10, 2020, during the first wave of the COVID-19 pandemic. The new law understands the informal caregiver as someone who provides assistance to another person in a situation of dependency, due to some type of disability. Such support could include, for example, the provision of food, transportation, support with hygiene and medication, as well as daily safeguards. Most of these caregivers are women who carry out this work exclusively, others are overloaded with the accumulation of tasks, in a—so often failed—attempt to reconcile professional life and care for other family members.

Since its legislation, the statute has been confronted with some problems of implementation. The statute, for example, theoretically provides support measures for informal caregivers and their referral to social support networks. But neither these referrals nor support measures have been defined, nor how they will be implemented. Although the definition of social support networks is attached to Ordinance No. 2/2020, there are still no regulations on providing information or referrals. As a consequence, there is no prediction about when and how support measures for informal caregivers will be implemented.

Another problem is the question of training referred to in Article 4, which establishes the generic requirements of the informal caregiver. However, it does not mandate any training on providing care for the recognition of the statute, neither prior nor simultaneous to assuming the role of caregiver. Whereas one would expect that training and qualification is required to guarantee the quality of correct and safe care provision to those in need, the statute only sees it as a measure of support that the informal caregiver can benefit from, rather than a necessity.

Another problem is that although the statute recognises a main informal caregiver may be entitled to unemployment benefits, it is contradictory with the specific requirement prohibiting any type of unemployment benefit from another source of work (Article 5, paragraph "d"). This situation implies that, in the event of cessation of work activity by the main informal caregiver with the right to unemployment benefits, the caregiver must relinquish her (most care givers are women) subsidy status to be approved. But there are more fundamental problems.

2.2 Political Problems of the Law

Some feminists[1] have welcomed the measure. It contains some of the demands from the Wages for Housework (WfH) campaign that spread across Italy, the UK and the USA in the early 1970s. The campaign was based on the claim that capitalism heavily depended on the unpaid reproductive labour of women, essential for the reproduction of the workforce in general. Their wageless work thus created the conditions for the reproduction of industrial society. This failure to recognise the economic value of women's work underlies two interrelated problems of capitalist/industrial modernity: the subordination of women to men and the subordination of reproductive care work to industrial production (Barca 2020, p. 28).

Despite the widespread political support for the measure, in this article, we assume a critical perspective on this legislation, which we categorise as a palliative, rather than a preventive policy of care. We understand that the promotion and approval of the Informal Caregiver Statute ultimately corresponds to legally accepting the State's lack of responsibility regarding healthcare and ensuring the quality of life for a growing, ageing and dependent population.

The transfer of care responsibilities, from society to the family, is profoundly reactionary. Behind the measure is the patriarchal logic of the family as a place for the reproduction of the workforce. Thus, in addition to having little effectiveness in actually implementing needed care and guaranteeing caregivers' rights, it fundamentally devalues the collective and socialised management of responsibilities and tasks of reproduction.

The law—like most ethics of care—appeals to a romanticised, supposedly affective character of housework and caretaking. Under romanticisation, we understand the use of a perspective that seems to portray something as better or more appealing than the reality. In her critique of the ethics of care, Stacy Clifford Simplican (2015) claims that care theorists like Eva Kittay tend to romanticise human capacities to maintain relationships. Kittay romanticises care when she depicts loving experiences of care, thus obscuring the everyday struggles and conflicts of carers and dependents. Thousands of people, especially women, live in despair and waste their lives, being hostages to the bourgeois State's lack of responsibility for dependent people. It is also relevant to emphasise that mostly poor workers, without the resources to hire care from private entities, are at the mercy of this romanticised bias.

For Amaia Pérez Orozco (2014), in order to break with what she calls a "reactionary care ethic," it is necessary to understand the capitalist logic that instils a culture of domestic work with a moralistic bias, detached from the real meaning of life:

> Femininity goes largely through a construction of itself for others, through, among other mechanisms, the performance of all residual jobs. Here, one of the keys that endow identity and social recognition with its own meaning is the realization of the tasks that make life possible, subjecting it to life itself. This logic operates primarily in the private-domestic sphere, where the responsibility for sustaining life when capitalist markets are located at the epicenter and attacks is subsumed (Pérez Orozco 2014, p. 168).

[1] Silvia Federici has been involved in the International Wages for Housework Campaign since 1972.

3 Capitalism, Patriarchy and Care

To start our critique of privatised care work, we should briefly inquire about the relation between capitalism and patriarchy. Friedrich Engels (2010) famously considered the establishment of the patriarchal family as intimately linked to the establishment of private property in sedentary societies. The patriarchal family established the supremacy of men over women in order to produce children of undisputed paternity; children who later become the heirs of their father's property. The legal constitution of women as patriarchal subjects and property culminated in modern capitalism and the State. In most modern civil codes, women (bourgeois, proletarian and peasant women alike) are legally made more or less dependent, placed in the guardianship of a husband or father. The modern law awarded the husband with controlling property, as well as the right of moral and physical dominance over women (Kollontai 1978). Deprived of public and citizens' rights, the woman is given a choice between the bondage of marriage or the embraces of prostitution—a trade despised and persecuted in public, but encouraged and supported in secret. Bourgeois society "crushes woman with its savage economic vice" (Kollontai 1978, p. 64), by restricting her activity to unpaid reproductive labour or paying very low rates through super-exploitation in the labour market.

A popular contemporary reference regarding the historical relation between accumulation of capital and gender is the work of Silvia Federici (2004). *In Caliban and the Witch*, she argues that the State-sponsored terror of the sixteenth- and seventeenth-century "Great Witch-Hunt" played an essential role in the patriarchal disciplining of women under modern capitalism. Like other Marxist authors, such as Maria Mies (1994), Harvey (2005) and others, Federici argues that the subjugation of women to the reproduction of the workforce was the third pillar of "primitive accumulation" or "accumulation by dispossession," along with the expropriation of workers and commons, and the enslavement of native Americans and Africans. Under this "primitive accumulation"—which is not "primitive" at all, but rather a continuing and contemporary logic of capital as Luxemburg famously asserted—we understand the continuing logic of use-value accumulation through violence at the capitalist system's frontiers (Van Vossole 2023).

Federici recognised the role of female procreative activity, particularly in relation to the population boom under capitalism since the nineteenth century, as a form of "primitive accumulation" rather than a simple consequence of economic development. According to Federici (2004, p. 91), procreation should not be approached as a "natural effect," but rather as "a social, historically determined activity, invested by diverse interests and power relations." She argues that the West-European women's destiny of enslavement for procreation was not dissimilar to that of female slaves in the American colonial plantations. The witch-hunts, as well as the subsequent maximisation of gender differences through feminine and masculine prototypes, were necessary to differentiate power between women and men, concealing women's unpaid labour under the guise of natural inferiority. Without this differentiation, capitalism could not have immensely expanded the "unpaid part of the working day" or used the (male) wage to accumulate women's labour.

Women's work, like that of peasants, the colonies or other "non-capitalist economies," as Rosa Luxemburg (2016) called them, is available as "free of charge" as nature. Under the capitalist patriarchy, the female gender is not the result of a congenital male misogyny, but a structural necessity for the process of capital-accumulation. A "never-ending process of capital accumulation, cannot be achieved unless patriarchal man-woman relations are maintained or newly created" (Mies 1994, p. 38).

3.1 Care and the Sexual Division of Labour

The result of this necessity is a sexual division of labour. Labour is reserved for (predominantly) men's productive work under capitalist conditions: meaning work for the production of surplus value. Women's contribution to the production and reproduction of life—care—has been literally defined as a function of their biology or "nature." Even the human body has been divided between the "human part," meaning the hands and brain, versus an "animal part," meaning the genitalia and womb. Much of women's productive activity has thus been "naturalised" and does not appear as labour, such as giving birth, raising children and domestic work. It does not appear as an activity interacting "with nature," but as an activity "of nature"—like the produce of plants, happening "unconsciously," without control by the involved subject. "Female productivity is animal fertility" like found in nature, so the subject finds itself in a relation of domination, with the activity usually unpaid or underpaid. The feminist approach to labour focusses on the whole of the invisible spheres of labour, particularly of unpaid labour, mostly done by women and/or racialised people (Dombroski 2020; Gibson 2020; Salleh 2004). Most of these forms of labour are care work. This includes care for people, for the natural environment, domestic labour and reproductive work in general to ensure the subsistence conditions of the community. This unpaid labour, like taking care of children or parents, cooking, sorting waste, etc., is an important part of social and environmental reproduction.

At the same time, it is important not to neglect social class in this division of labour. Feminist questioning of poverty has been a powerful ally in the conceptual elaboration of a political understanding, that allows breaking with an economist view of poverty as only the absence of income or lack of access to consumption, while showing the gender dynamics behind its recreation. The Brazilian philosopher Suely Carneiro (2003, p. 50) brings an important reflection on the hegemonic assessment regarding the dynamics of oppression and power for precarious women: "When we talk about the myth of female fragility, which historically justified the paternalistic protection of men over women, what women are we talking about?" Certainly not those that have to combine a stressful underpaid job with the care of children, the elderly and other household tasks.

Most precarious caregivers living in Portugal would never recognise themselves in the fragility myth, particularly those in the most fragile position, like peasants, women of colour, etc. The justification is simple: these women were never treated as fragile. Rather, they are part of a large number of precarious workers who have probably never understood when certain feminist discourses suggested or told women that they should take to the streets and work. The bodies of these women have always been subject to exhausting work and a condition of subalternity.

As a consequence, a majority of women do not recognise themselves in certain liberal feminist speeches which emphasise overcoming the romanticism that permeates images of beautiful delicate young women who inspire the poetic fado music of elitist university cities like Coimbra. A huge contingent of women has never been muses of anything, on the contrary.

Therefore, rethinking care policies involves the need to democratise the feminist movement. Concretely, this change is only possible by establishing and incorporating guidelines that consider the impact of class and how violence disproportionately affects a significant portion of the female population, particularly those who are not daughters of the elite. Given that capitalism is characterised by precariousness, undervaluation, and the unrecognition of labour, it is crucial for working-class women to organise politically, strengthen bonds and build bridges towards alternative horizons within social reproduction.

3.2 Towards the End of Care?

Female domestic work was not always what it is today. Like all social phenomena, it is a fluid product of human history. In *Women, Race and Class*, Angela Davis (1983, p. 223) asks: "If it were at all possible simultaneously to liquidate the idea that housework is women's work and to redistribute it equally to men and women alike, would this constitute a satisfactory solution?" Although many women welcomed the advent of the "household man," the "desexualisation" of housework is unable to change the oppressive nature of work itself within the capitalist system. Ultimately, Davis concludes that neither women nor men should waste precious hours of their lives on work they find neither stimulating nor productive.

The rapid expansion of the female workforce has meant that more and more women found it difficult to continue exercising their role as free caregivers according to traditional patterns of exploitation. The "industrialisation" of domestic work, together with its socialisation, became an objective social necessity—albeit not immune to capitalist alienation. At times, the task of care as an individual private responsibility of women, carried out under primitive technical conditions, seemed to approach historical obsolescence. Nevertheless, it has prevailed in social attitudes that continue to associate the female condition with social reproduction, domestic work and care for family members.

A substantial part of the tasks inherent in housework and care work could be—and often already is—actually incorporated into the public and private economy. Angela Davis (1983), in the light of Marxist theory, argues that domestic work no longer needs to be considered necessary and unalterably private, particularly given the technological advances that could radically transform the nature of housework. Teams of trained and fairly paid workers, shifting between residences, using technologically advanced means for cleaning, could quickly and efficiently accomplish what today's "housewives" do in an arduous and primitive way. The same logic applies to other aspects of healthcare, guaranteeing both qualified jobs and working conditions to trained, competent professionals, as well as a more dignified life for the sick person and their family.

The question then arises: why is there such a lack of interest in radically redefining the nature of domestic work and care? The answer probably lies in the fact that the capitalist economy seems to be structurally hostile to the "industrialisation" of domestic work. The primary reason is that maintaining the logics of accumulation by dispossession, of violently robbing free labour (similar to what occurred for so long through the maintenance of colonies and slavery) contributes to a free increase in profits. Eco-feminist scholar Ariel Saleh (2019) has called care tasks—such as childcare, elder care, forest nurture—a "meta-industrial" labour. This type of labour produces no exchange value, but instead a "metabolic value form" that flows into and sustains the essential bio-infrastructure of the capitalist system. She argues that by global population numbers, meta-industrials make up the global majority of workers.

Another reason is because people dependent on care—excluding those rich enough to hire full-time servants—are usually not the wealthiest customers. A commercialisation of most care work would result in little profit. Like all unprofitable businesses, it forms an anathema to the capitalist economy. Often, it is the government that needs to intervene in such cases. But if care work is socialised, this usually implies the need for large government investments to guarantee access to working-class families, whose need for such services is more evident.

4 Socialised Public Care

The welfare state model, which was only fully implemented in Western Europe after the Second World War, did socialise some aspects of care work. The pillars of that welfare state included some aspects that, for a long time, had been a family affair or voluntary religious works. Governments started to guarantee the right to healthcare, through the implementation of national health services and the expansion of hospitals, which not only relieved many women from tasks of caring for sick family members at home, but also created an enormous increase in the quality of care through massification and scientific advances. The same holds for the right to education, when a massive expansion of the public education system guaranteed education, as well as food and entertainment, from nurseries to universities. The welfare state would go on to guarantee social protection—including the set of benefits that cover risks from not having a job—and assistance in some situations of dependency, with the creation of public elderly homes, etc.

Over the years, all have been subject to criticism from a feminist point of view, on issues of applicability and effectiveness. The lack of applicability and effectivity is often not recognised because of the presumption that women and families still fill in the gaps. By assuming women belong to such a family role, they are prevented from sharing the same conditions as men in the labour market. For Amaia Pérez Orozco (2014, p. 130), it is necessary to demand that the State assumes responsibility for what has been called the "fourth pillar of the welfare state": assistance in situations of dependency.

From a feminist perspective, we need to question the extent to which the welfare state protects against the risk of decreasing well-being levels when the unpaid care work—historically assigned to women—ceases to exist. We are talking about the demise of the family. To what extent can (women in) families stop doing their free care tasks without people who need care losing quality of life? This aspect can be a starting point for interpreting welfare systems as a conjunction of State-market-home spheres that provide resources. Depending on which priority is given, we can say that well-being is more or less (de)commodified and (de)familiarised (Pérez Orozco, 2014, p. 131).

Health and education systems do not cover full care needs related to health and illness, nor childhood. A large part (in fact, most of it) is covered by families. Across the world, women and girls are performing more than three-quarters of the total amount of unpaid care work (ILO 2008, p. v). Likewise, it is recognised that hospitals would not be able to handle all the labour inherent in care if there were no relatives of hospitalised people to provide home support. The situation is considerably worse in cases of precarious patients for whom the State denies responsibility to feed, clothe or even provide medication.

The welfare state was conceived as the express attempt to build society under the tension between capital and life, in which the State can establish norms that redistribute what capitalist markets barely distribute, guided by principles of solidarity, recognised as universal rights. However, these measures do not eliminate the fact that the welfare state is built upon a sexual division of labour. Taking the capital-life conflict seriously implies recognising that capitalist markets are central to the debate and that an extensive lack of collective structures persists for guaranteeing life-sustaining processes. This underlies the various models and their variable conjunctions with capitalism and heteropatriarchy—also the welfare state.

The responsibilities assumed by the welfare state in this regard are not only partial and incomplete, but also extremely fragile, being called into question during times of heightened tension, as seen in the current crisis. Positioning the markets at the epicentre inhibits the existence of a collective responsibility when setting conditions for the possibility of life. At the same time, life must go on. How and where? This responsibility is transferred to networks of intimacy (households), with privately available means, which fundamentally consist of two: unpaid work and money market consumption. Responsibility is privatised because it remains inserted in the private-domestic sphere, therein becoming invisible and feminised (Pérez Orozco 2014, p. 132).

This is why we believe that the sexual division of labour continues to exist in these invisible and feminised jobs, inherent to the welfare state. Everything that is built on the capital-life conflict is also based on the sexual division of labour. Living under capitalism means being subject to the process of accumulation. Maintaining the welfare state means maintaining this mechanism. Our own ethical and political conception of life is colonised by capitalism, limiting our interventive priorities in the game of accumulation, without going beyond the welfare state. Understanding these dynamics of oppression is a crucial strategy for us to resignify care as a component that generates, not imprisons, life.

To achieve this, it is necessary to go beyond creating metanarratives that explain history in a favourable way for women. Instead, it is essential to recognise that many culturally denied aspects, if explored, could allow for a radically different understanding of the one that has been told. This history must be necessarily situated and created collectively, putting partial truths in dialogue in the search for collective cartographies. In this sense, creating knowledge from hegemonic positions does not serve us, as they impose themselves as the only possible truth and prevent the transformation of reality.

4.1 Care in Crisis

Feminist theorists have long considered capital accumulation by dispossession as a process fundamental and socially guaranteed by capitalism. The phenomenon becomes even more flagrant in times of crisis, when markets need to expand beyond their existing limits in the entire socioeconomic structure to overcome contradictions and compensate for losses. Since the establishment of the welfare state, the crises of capitalism have often been expressed through transfer mechanisms in a fiscal and or social crisis (Habermas 1975, O'Connor 2002). The crisis affects care activities in both public and private dimensions. A state undergoing a fiscal crisis tends to cut crucial social services to reduce costs. For instance, the Euro-crisis following the Global Financial Crisis and the ensuing austerity measures led to reduced public spending on education, healthcare, social benefits, etc. When the state fails to provide access to care services, meeting unavoidable needs quickly becomes unpaid work. These unpaid jobs increase, replacing the public services that the state no longer provides.

Additionally, the degradation of living conditions becomes more evident, negatively impacting health and increasing care requirements, while also worsening conditions for domestic work. This increase is evident from cuts in public spending on dependent care and childcare. The scope of this increase in unpaid work is extensive: from spending thousands of hours cooking cheaper and less elaborate meals, to the time spent looking for and offering palliative healthcare.

According to Antonella Picchio (1999), in the context of crisis, the subsistence strategies of many families suggest that well-being does not depend exclusively on the situation of its members in the labour market. Their well-being also hinges on a complex and diverse network of activities that occurs in an unaccounted number of spaces and times. While production is globalised, the reproduction of the working population remains local. This makes it easier to hide the growing tensions between production and reproduction, "[...] even when their effects are dramatically visible" (Picchio 1999, p. 225).

4.2 Care in the Pandemic

Rarely has care been so central in a crisis as in the case of the recent global coronavirus pandemic. Many authors have argued that the COVID-19 crisis is not an external shock, but a crisis of capitalism itself. The crisis not only originated from the increased pressure and interaction between society and nature due to capitalist expansionism, with the transfer of pathogens between wild animals and the human food chain. The risks of contagion were also multiplied by industrial meat-production and the concentration in urban centres, while the speed and global reach were thanks to the capitalist globalisation of trade and transport.

As a consequence of the pandemic, care came to occupy a central position. Healthcare systems and hospital services crashed due to the peaks of sickness and deaths provoked by the pandemic. Economic and social systems were disrupted by governments and protest movements alike, in a global attempt to protect and care for life. At the same time, people had to stay at home, which meant a huge increase in household care labour, as schools and care institutes were sealed from the outside world. For this reason, philosopher Silvia Federici (2020) points out that the advent of the pandemic reinforces the need to rescue the sphere of labour connected to life.

In early 2020, Oxfam published a report entitled "Time to care." It centred the discussion on unpaid and underpaid care work and the global crisis of inequality, the demand for public policies and actions to strengthen new forms of feminist economics and care that operate outside capitalist logic. The report scrutinised the gap between rich and poor, highlighting how the contemporary economic system plays an important role in producing these inequalities.

Social reproduction consolidated as a fundamental ground of dispute. While the rest of societies collapsed, the public care services appeared as the fundamental core of society. These essential services had been underfunded over the decades, with the unpaid invisibilised work in households or by marginalised and often racialised precarious workers.

5 Alternatives from Crisis

The first "alternative" seemed to continue at full throttle on the paved neoliberal road. Governments often refused to intervene in industrial production, while outsourcing fundamental services for the public management of the pandemic. It suffices to mention how the vaccination programs were totally dependent on private pharmaceutical companies, with intellectual property (patents) being prioritised over free access for poor countries. The objective need made care prone to profiting certain individuals and large corporations. Most of the political and economic measures that were enacted in response to the pandemic crisis have this objective.

Ironically, the efforts to create an Informal Care Statute, like in Portugal, could be included in this drive to further privatise the remnants of the welfare state in a period of crisis. In *Portugal: A Singular Portrait*, Portuguese sociologist Boaventura de Sousa Santos (1993) explains the country's the permanence on the periphery of Europe as being the equivalent of a Global South in the Global North. This position of ambiguity highlights both Southern and Northern epistemologies. Wage relations and social well-being in semi-peripheral regions like Portugal explain many of the contradictions and intersections between the hyperlocal and the transnational. This reflects a combination of the World-System dynamics with regulations applied to local realities. The Portuguese Welfare State was only implemented after the revolution of 1974, a moment when the rest of Europe's welfare states had already entered a downturn. Public care was, therefore, always less effective than recognised in the constitution. Santos attributes a certain "flexibility" of care in Portugal during periods of crisis, as informal social networks and families—particularly in the countryside—tend compensate for State failure. Also now, at a moment when States refrained from investing in public institutions of care, the Informal Care Statute institutionalises a privatisation of care practices in the family sphere, which had already been present.

These contradictions go a long way towards explaining the maintenance and acceptance of public policies that are still so "rudimentary" in terms of domestic work and the perspective regarding the sphere of care as a whole. The management and applicability of Law No. 100/2019, establishing the Informal Caregiver Statute, makes us reflect on how the romanticisation of care is responsible for maintaining women in caregiver roles, while removing the State's responsibility of providing citizens the means to have a more dignified life. Rather than romanticising this position as a form of solidarity, a critical perspective should recognise this as a return to a peripheral condition of primitive accumulation.

What is the alternative then? Angela Davis proposes a full socialisation of care work as the only way for an anti-capitalist emancipation of women. For Davis (1983, p. 243):

> The abolition of housework as the private responsibility of individual women is clearly a strategic goal of women's liberation. But the socialisation of housework—including meal preparation and child care—presupposes an end to the profit-motive's reign over the economy. The only significant steps toward ending domestic slavery have in fact been taken in the existing socialist countries. Working women, therefore, have a special and vital interest in the struggle for socialism. Moreover, under capitalism, campaigns for jobs on an equal basis with men, combined with movements for institutions such as subsidized public health care, contain an explosive revolutionary potential. This strategy calls into question the validity of monopoly capitalism and must ultimately point in the direction of socialism.

Based on Davis's reasoning, the demand for a truly socialist care economy can only be conceived through: nationalisation of the entire private network of homes and institutions that care for dependent people; immediate nationalisation of the entire network of public-private (often church-related) care sector; extinction of orders of health professionals, which make it difficult to allocate human resources within the health sector due to their corporatist nature; creation of national training programs for geriatric assistants and similar professions; and unionisation of workers in homes and other care institutions, etc. Society should provide free collective cleaning services, collective canteens, nurses that accompany patients at home, nurseries etc., while offering quality jobs and decent salaries for the people working in those care services—equivalent to other economic sectors in society.

Only when the vast majority of care work is socialised and the remaining care work is done out of love, rather than necessity, can we achieve true emancipation for women and establish free and healthy relationships with those who currently depend on others. Care for others should be approached in the same way as those who cook at home, as a hobby rather than need. Only then can one abolish the alienation inherent to care work.

6 Conclusions

Throughout this article, we have analysed the concept of informal care, based on the practical example of legally institutionalising informal care in Portugal at the onset of the COVID-19 pandemic. Despite the very progressive reasoning behind the law's implementation, a short analysis brought forward some practical problems in terms of implementation and effectiveness, as well as some more fundamental theoretical problems. We have argued from a feminist perspective that the implementation of the Informal Caregiver Statute in Portugal should be seen as the institutionalisation of inadequate State intervention; effectively privatising care work and reproducing the already patriarchal sexual division of labour. We have argued that under capitalism, care work in households is a form of primitive accumulation, or accumulation by dispossession of unpaid and invisible labour done by women. This kind of dispossession is stronger in (semi-) peripheral societies with deficient welfare systems, especially in times of economic and social crisis.

The attempt to reform capitalism and make it more humane has failed to provide us any guarantee that inequalities will disappear. On the contrary, the repeated assistance and emergency measures in cases of crisis and extreme scarcity have alienated and silenced struggles for alternatives, leaving a relevant gap in the fight for social and economic justice for all. If we want a structural, human alternative to care that can overcome the sexual division of labour, it is necessary to overcome the capitalist system. Capitalism alienates us from a true policy of care and inserts women into a system that depletes women's emotional energies, bringing the alienation of the market economy within family relationships, between dependent people and their caregivers.

References

Barca, Stefania. 2020. *Forces of reproduction: Notes for a counter-hegemonic anthropocene*. 1st ed. Cambridge University Press. https://doi.org/10.1017/9781108878371.

Carneiro, Sueli. 2003. Enegrecer o feminismo: a situação da mulher negra na América Latina a partir de uma perspectiva de gênero. *Racismos contemporâneos. Rio de Janeiro: Takano Editora* 49: 49–58.

Davis, Angela Y. 1983. *Women, race, & class*. Vintage.

Dombroski, Kelly. 2020. Caring labour: Redistributing care work. In *The handbook of diverse economies*. Edward Elgar.

Engels, Friedrich. 2010. *The origin of the family, private property and the state*. Reissue edition. London: Penguin Classics.

Federici, Silvia. 2004. *Caliban and the witch: Women, the body and primitive accumulation*. New York: Autonomedia.

———. 2020. Revolution at point zero: Housework, reproduction, and feminist struggle. PM press.

Gibson, Katherine. 2020. Collectively performed reciprocal labour: Reading for possibility. In *The handbook of diverse economies*. Edward Elgar.

Habermas, Jürgen. 1975. *Legitimation crisis*. Beacon Press.

Harvey, David. 2005. *The new imperialism*. Revised edition. Oxford/New York: Oxford University Press.

ILO, International Labor Organization. 2008. *Care work and care jobs for the future of decent work*. Geneva.

Kollontaï, Alexandra. (1978). *Selected writings of Alexandra Kollontai* (Trans. A. Holt). Lawrence Hill.

Luxemburg, Rosa. 2016. The accumulation of capital: A contribution to the economic theory of imperialism. In *The complete works of Rosa Luxemburg, volume II: Economic writings 2*, ed. Peter Hudis and Paul Le Blanc . New York: Verso.Reprint edition

Mies, Maria. 1994. *Patriarchy and accumulation on a world scale: Women in the international division of labour*. 5th ed. London/Atlantic Highlands: Third World Books/Zed Books. Distributed in the U.S.A. and Canada by Humanities Press.

O'Connor, James. 2002. *The fiscal crisis of the state*. New Brunswick: Transaction.

Orozco, Amaia Pérez. 2014. *Subversión feminista de la economía: aportes para un debate sobre el conflicto capital-vida*. Traficantes de sueños Madrid.

OXFAM – Brasil. 2020. Tempo de cuidar. *O trabalho de cuidado não remunerado e mal pago*.

Picchio, Antonella. 1999. Visibilidad analítica y política del trabajo de reproducción social. In *Mujeres y economía: nuevas perspectivas para viejos y nuevos problemas*, 201–244. Icaria.

Saleh, A. 2004. Sustainability and meta-industrial labour: Building a synergistic politics. *The Commoner* 9: 1–13.

Salleh, Ariel. 2019. Planetizing the labor movement. *Great Transition Initiative*.

Santos, Boaventura de Sousa. 1993. *Portugal, um retrato singular*. Vol. 5. Edições Afrontamento.

Van Vossole, Jonas. 2023. *Marx's primitive accumulation as a perspective on nature, race and gender*. Preprint. SocArXiv. 10.31235/osf.io/nrbgp.

Chapter 12
On Caring Acts and Technological Mediation Processes

Joaquim Braga

Abstract In this paper, I explore whether caring acts can be properly understood through technological mediation, as seen in telehealth and telemedicine. It aims to challenge the reduction of such acts to instrumental criteria and exclusively utilitarian ethical principles. I will determine whether the relational dimension between caregiver and care receiver, which is fundamental in caring acts according to the ethics of care, remains unaltered in technologically mediated acts, or if these acts can entirely supplant it. In conceptual terms, the paper will emphasize the concepts of "relation" as the core of caring acts and "mediation" as the framework for understanding the nature of technology.

Keywords Devices · Mediation · Medical care · Reciprocity · Relation

1 Introduction

When we think about the technological advancements of recent decades, the widespread integration of automation principles into the design of new artefacts and systems is hard to ignore. Although not tangible entities, such principles reinforce the imaginary that allows the human mind to project itself into the utopia of its operative self-sufficiency. Recognizing this imaginary background and its social and cultural implications, an ethics of care tends to take a different approach. Current research on technology in care, particularly in medical settings, generally aligns with two theoretical approaches employed in conceiving the technological world: the deterministic theory and the neutrality theory. The authors who support the latter believe that the benefits of technology depend on its application in care settings; the authors who support the former argue that technology can compromise the humanization of medical procedures by reducing them to mere instruments.

J. Braga (✉)
Department of Philosophy, Communication and Information, University of Coimbra, Coimbra, Portugal

J. Braga (ed.), *Ethics of Care*, Advancing Global Bioethics 20,
https://doi.org/10.1007/978-3-031-73608-7_12

Since they involve vital needs, medical acts still function today as a reference for acts of care. To what extent, however, can they serve as a subject matter for the theoretical articulation of technological mediation and the broad theme of care? Is the context and nature of institutionalized care not being overlooked? The idea that the relationship between caregiver and care receiver can be thought of in the light of an ideal cognitive locus—following the Cartesian epistemological model of a subject facing an object—is, at all, equivocal and fallacious. One can never forget that, for instance, in many hospital settings, due to the insufficient number of professionals and the intricate modes of organization, there is an empirical lack of care relationships encompassing affectivity, trust and reciprocity. Nevertheless, it is certainly in the field of medicine that a large part of the questions about the role of technology in the interaction between caregiver and care receiver have arisen. For instance, in his article The Physician's Role in a World of Technology, Joel D. Howell asserts that "The question, 'Do patients want a competent physician or a caring physician?' is meaningless. The caring is part of the competence. Ultimately patients want a physician who really, truly 'gives a damn' about what will happen to them. Physicians have very personal relationships with their patients" (Howell 1999, p. 247). The article ends with Howell expressing a "desire", rather than "confidence", regarding this matter: "That close relationship between patient and physician must persist if new technologies are to bring us closer to a utopia, not a dystopia, and if the good of individual patients is to endure as a central value of health care" (Howell 1999, 247).

In general, the relational nature inherent to the acts of caring does not presuppose the distinction of those that, for their accomplishment, imply technological mediation processes. Regardless of the outcome, caring always begins with the idea of closeness between the caregiver and the care receiver as the prime root of relations involved in care. However, since caring lies in mutual interaction, it's pivotal to understand how technological intervention impacts it. To this theoretical end, I start from the semantic principle that, while not mutually exclusive, "mediation" and "relation" are not interchangeable concepts. The inherent instrumentality of mediation alone cannot maintain the mutual nature implied and derived from caring relations.

The question of knowing how to articulate an instrumental approach to caregiving with a humanistic approach is also at stake here. But for this to be feasible, it is crucial to refute deterministic and technology-neutral theories. Two main topics will be addressed in this chapter. The first concerns the reproduction of care as a reciprocity-based human practice. The second—on which the first depends a lot—gets on developing an awareness of technological mediation and the active role that caring acts play in shaping it.

2 Hypermediation and Modern Society

Modern society is in large part of its spheres—labour, commercial, informational, educational, political—greatly influenced by a hypermediation phenomenon. Several technological inventions, allied to their digital convergence, have significantly increased the reliance on devices (the possibilities and limits imposed by them) for human activities. The theoretical implications of this are very evident in media theory and the philosophy of technology. For example, in his extensive work on technology, Don Ihde revises Emmanuel Levinas' concept of otherness to apply it to certain technological relations between the user, the artefact and the world. Ihde aims to demonstrate how technologies "become other or, at least, quasi-other" (Ihde 1990, 98). Technological quasi-alterity points to a latent anthropomorphism in human interactions with artefacts in daily tasks, renewing the social imaginary of full automation. In this sense, Ihde acknowledges that there is already anthropomorphism in the excessive attention given to an artefact someone owns (Ihde 1990, 98). Additionally, Ihde always regarded techno-fantasies as ways to surpass the limits of the technology itself. Now, the alleged alterity relations of technology and the transfer of the idea of care to the domains of artefacts are two symptoms accentuated by the hypermediation phenomenon. In both cases, there is something paradigmatic in common: the concept of relation lacks a focus on reciprocity.

When considering care from a technological perspective, we shouldn't limit it to specific care practices involving technological tools. Regardless of such practices, technology shapes our understanding and promotion of care on an individual and social level. With the emergence of complex devices, calling for a type of care for the machine is becoming commonplace. Establishing a relationship with the machine becomes a prerequisite for using it. In such a tendency, there is an emotional projection of human care onto something incapable of caring for. A second significant tendency arises when an alleged capacity for care is projected onto the machine itself. Indeed, the two tendencies are interconnected and influence each other, shaping technological innovation.

Given these anthropomorphic projections, it's crucial to critically examine the concept of care within the contemporary technological framework. Are we truly discussing care when it's projected onto artefacts, or is it just a misuse of the term?

The widespread use of technology in our daily lives has led to a semantic regression in the meaning of "caring" and "care". Now, any concern or service provision is often labelled as care. However, being concerned about something doesn't necessarily equate to taking care for someone—this is the first premise of a critical understanding of care. Unlike concern, care involves a mutual bond among individuals. The second premise goes beyond semantics and touches on an ethical dimension. The concept must not only define care, but also help identify genuine care practices.

The decrease in reciprocity is a structural condition in healthcare domains due to technological advances and the scientific applicability of the artefacts themselves. Finn Olesen argues that with the spread of stethoscopy, two fundamental differences are introduced in healthcare: in therapeutic terms, "medical science came to be more concerned with diseases than with patients", and in epistemological terms, "the patient's knowledge is particular and subjective while instrument-driven knowledge is general and objective" (Olesen 2006, 236). However, in this technological specialization process of medical knowledge, another relevant fact should be considered. By creating a distance between doctor and patient, the emphasis on objectivity over reciprocity promotes the formation of a cultural belief in the reliability of technological devices for personal use and the ability of any individual to use them without professional mediation. Today, numerous self-tracking devices and the information and communication processes they support exemplify this tendency (Braga 2021).

One of the most significant effects of technological advancement in modern society is the widespread perception that interpersonal relationships can be easily *replaced* by technologically mediated relationships. For example, Mark Coeckelbergh discusses the delegation of care in various social areas, where care has shifted from "family, neighbors, and friends to professionals and institutions", and poses the question: "Why not delegate care to robots?" (Coeckelbergh 2012, 281). He suggests that robotic devices can provide "companionship" to care receivers. Although claiming that "the end of companionship cannot be reduced to pleasure or utility" (Coeckelbergh 2012, 285), he overlooks that robotic programming requires algorithms to quantify the benefits, individually and socially. This approach cannot surpass Jeremy Bentham's hedonistic ethics and his utilitarian calculus ((*felicific calculus*), which is based on the quantification of pleasure. The utilitarian calculus not only determines the moral nature of actions by quantifying pleasure and pain but also establishes a strict distinction between what causes pleasure and what causes pain.

The "Environmental Hypothesis" proposed by Darian Meacham and Matthew Studley (2017, 99) suggests a radical idea of caring machines. Starting from an equivocal cognitive separation between psychic states and expressive processes, they argue that robotic devices could create medical care environments as they believe these "are wholly contained or made manifest in external movement" (Meacham and Studley 2017, 99). What remains from the environmental hypothesis is, strictly speaking, as the authors assert, the alleged idea that "when a generation that has grown up with robots as playthings, assistants, and perhaps colleagues reaches old age, its capacities and potentials for human-machine relations will be greater than what studies today show" (Meacham and Studley 2017, 109). In other words, the authors appeal to a kind of human behaviour self-programming, regardless of the interaction nature that may exist between humans and machines. However, alterity relations are triggered and permeated by an *expression of life*—the lives of beings or things that enable and evoke the lives of beings. Such is not the case with robotic devices, mainly because they are intended to be simulacra and, consequently, substitutes for living beings and not just mere tools.

The technomyth of substitution is reinforced by the concept of artificial intelligence and post-humanist theories, where intelligence is reduced to a *problem-solving* idea. The attempt to humanize robotic systems and use them in care provision has led to the employment of the concept of relation to human-machine interaction. Nevertheless, this relationality lacks reciprocity and, in a language based on Alan Turing's computation theory, "intelligent" machines simultaneously acquire the status of *caring machines*. The fusion of the technomyth of automation with that of substitution—along with the traditional substantialist ontology of the mind and body—has led to the perception of new technological devices as intelligent machines with relational capabilities. This prompts the question of if the claimed relationality can be algorithmically programmed to recognize and solve problems accurately.

In a purely instrumental approach, one could accept the thesis that technology always supplies caregiving with the argument that it also supports problem-solving. Taking that argument to its logical conclusion, when issues arise, be they therapeutic or otherwise, there's a chance they can be resolved with the aid of artefacts. But what is wrong with this type of argument? Above all, the fallacious judgment lies in the viewpoint that using technology equates to identifying the root of problems. We can never know if drawing attention to problems is entirely triggered by the effectiveness of their resolution. That is to say, *cure* by itself does not guarantee *care*. There is something more to what we call "care"—and that *something more* is not only guaranteed by the artefacts used to undertake it. When caregiving is reduced to isolated acts of problem-solving, it doesn't mean that care will continue to be a fundamental human behaviour.

3 Relation and Mediation

The axiological neutrality thesis stands for technology in the same manner consequentialism stands for ethics. Both rely on the instrumental relationship between means and ends as their fundamental criteria. Now, it is precisely this reduction of ethics to instrumentality that contradicts the aims of an ethics of care. An act of care involves an anticipation gesture, setting it apart from simply solving a problem. A relational ontology is grounded on the embodiment of otherness in every person's life, which arises from the existential primacy of care over concern. Thus, caring in its relational sense implies positioning every care receiver in a generative interaction context, assuming that his/her individuation can serve as a model for others' individuation. The anticipation gesture in our daily lives means that care extends beyond the therapeutic domain to include precautionary feelings and actions.

The doctor-patient therapeutic model, which now includes telemedicine, is not enough to define care and caring. This is because vulnerable people do not always express their condition in words or seek help from others. Thanks to technological advancements, modern scientific research is constantly finding new ways to quantify and measure things. Nevertheless, twentieth-century philosophical thought aimed to move away from this goal and demonstrate that the human body cannot be fully understood or captured through mere calculation and measurement techniques. Martin Heidegger (1987) and Hans-Georg Gadamer (1993), among others, strongly opposed the modern trend to quantify the human body, reducing medicine to a purely quantitative science.

The example of telemedicine illustrates the shift in proximity relations caused by hypermediation. Bringing a patient's body, which is physically distant, closer through technology also creates a predetermined distance between the patient and the doctor. Supporters of telemedicine argue that it improves our health system. Notwithstanding all claims of instrumental efficacy, the critical question is if better health truly means better care.

At this point, it is advisable to introduce a distinction between care and cure, as suggested by Sally Gadow. In her humanistic approach, care goes beyond healing and encompasses a commitment to the vulnerability of others. Cure procedures and those of care are asymmetrical because the former "achieve their effect through the exercise of power by one person over another"; such kind of power tends to redouble "the vulnerability of the one over whom it is exercised, no matter what benevolent purposes the power serves". Care, in the narrow sense, can mitigate "the overpowering of one person by another that cure entails" (Gadow 1988, 7). Gadow argues that technologically mediated medical care can shift the doctor-patient alterity relation to a purely technological interaction between user and artefact. In other words, such a shift can reduce the human aspect of care to a focus on operability.

Raising awareness of vulnerability is crucial for the anticipatory gesture of care to be a part of human perception and everyday sociability. Not making time for others means, in effect, not receiving the time articulated by caring for others. Without mutual relationships, we lose a sense of time tied to vulnerability. Compassion, for instance, being an intrinsically social feeling, exerts a retroactive effect on the way we deal, individually, with our psychic states. In today's context of ecological concerns regarding the planet's biodiversity and sustainability, the effectiveness of collective actions relies on the aesthetic rejuvenation of the sensible and sensibility, and on acknowledging others' bodies as extensions of our own. There is an urge to create social and institutional contexts that encourage caring practices and focus attention on what brings us together. By taking a holistic view of humans concerning the environment, we can develop new social awareness of vulnerability that is not just represented but genuinely embodied.

Accordingly, to understand the concept of care, we must consider that caring involves embodying life and well-being as a daily gesture. Repeating and spreading that gesture is easier than learning a new language. Mimetic behaviour plays a significant role in this process, as our interactions with other beings, whether voluntary or not, often trigger our caring gestures. That is why a philosophy of care focuses on practices rather than just values. Due to increasing technological mediation, our opportunities for these types of interactions are decreasing. The effects of technology use and control significantly interfere with the spontaneous ways in which we perceive, feel and act in the world. Consequently, the focus here is more on mediation than on relations. In general, due to the nature of our lifestyle, our behaviour is often limited to specific problems and individual concerns. However, it's up to us to make care a consistent part of our actions and institutions.

4 Presence and Caregiving

Articulating caring acts with technological mediation processes displays relevant possibilities, forming a consciousness of mediation. The need for a balance between the two transcends ethical imperatives and the requirement that instrumentality never overlaps humanity. Along with this balance comes the understanding of how technology is shaped and for what purpose it is created. The assumption that such possibilities inherent to the knowledge of the nature of technology can be reached through caring acts also suggests that they offer an experience of time that differs from other daily activities. As I have argued elsewhere (Braga 2021, 133), care time has a double meaning related to what needs to be taken care for in the present and what can be prevented from needing care in the future.

Based on efficiency principles and the operative nature of its artefacts, modern technological mediation mechanizes and accelerates temporal dynamics, structuring time as reaction time. Taking time to care for others involves mitigating the acceleration of temporal dynamics and creating mental and practical conditions favourable to feelings, thoughts, and reflection. This is especially evident in parent-child relationships, where care merges with education.

According to many empirical studies, increased technological alienation from parents can result in behaviours not in line with face-to-face interaction. Parents' excessive focus on mobile devices in the presence of their children can trigger various reactions, as reported by Radesky (2014, 847):

> caregivers absorbed in devices frequently ignored the child's behavior for a while and then reacted with a scolding tone of voice, gave repeated instructions in a somewhat robotic manner (eg, without looking at the child or relevant to child behavior), seemed insensitive to the child's expressed needs, or used physical responses (eg, one female adult kicked a child's foot under the table; another female caregiver pushed a young boy's hands away when he was trying to repeatedly lift her face up from looking at a tablet screen).

What is most relevant in these empirical observations is how face-to-face behaviour is easily changed by the sensibility aroused by technological means, such as when visual contact with others—the children, in this case—gets influenced by the device screen.

Excessive exposure to screens results in rapid and unmeasurable shifts in how humans perceive and internalize themselves since the visual is favoured by the senses in the self-image-making processes. With digital exposure and communicative pressure, the structuring role of perception in social relations is limited to the domain of information transmission. The belief that, regardless of their material nature, all occurrences are capable of technological mediation shapes the *technomyth of substitution*. Thanks to current technological development, the technomyth of substitution reinforced its inscription power in social imaginary, even though feelings related to distance do not equal those stimulated by proximity.

The concepts of cure and care, although not exclusive, are not synonymous with those of presence and attention. One will always assume that face-to-face interaction can be replaced by remote interaction if one takes for granted the "risk and opportunity" argument (Sen 2016, 1069). Such is the case when one accepts the widespread principle that online relationship forms "appear similar to those which marked relationships in a pre-digital age"; as well as the argument based on it that "the solidity of social relationships – for good and bad – had not melted away as fundamentally as some accounts have claimed" (Sen 2016, 1072).

Being not a matter of demonizing it but bearing in mind the changes introduced in human relationships, it becomes imperative to look back at how we became aware of technological mediation. Several studies on the use of mobile devices in the relationships between parents and children only focus on the positive and negative transformations triggered by them (Kildare and Middlemiss 2017). Despite their relevance, it is a misstep to limit issues of technological mediation of care to purely ethical considerations. Above all, it also remains to inquire about the impact of such changes on the coming-into-being of technological literacy. With the increasing dissemination of digital technologies and their inscription power in most daily life areas, there tends to be a parallel *regression* of the ontological status of media. Media are more than objects. Rather, they are only true media because they transpose the epistemological sphere of object perception and allow the articulation and awakening of intersubjective links in human experience. Parents who give their children a mobile phone to play with are promoting regression of the medium, which is its regression to a mere object function. The formation of an awareness of technological mediation will be jeopardized by this fact, which will also change the nature of human interaction.

5 Final Remarks

Overvaluing human agency is a common feature of our modern technological imaginary. It is an integral part of technology to connect with the realm of the possible, with all those goals that can be achievable. Technology makes us look at the future—to the possibilities that can come into being if we use its artefacts. Technology promises to keep care practices at the core of our daily lives. The possibilities of technological mediation have always been visible, from medical care to so-called informal care. The goal might shift, but the technological inscription power is still being reinforced in multiple ways. In the seventeenth century, the thermometer was announced by the Italian physiologist and physician Santorio Santori as a means capable of serving medical and individual use. Likewise, nowadays, several engineers, scientists and thinkers present robotic devices as a reliable solution for medical and informal care procedures.

The idea we build and reproduce about care is determined by the continuous expansion of technology and its view of care practices. It is not enough to have a pure instrumentalist approach to technology to understand the notion of care it develops. The technological imaginary needs, more than ever, to evoke care to get constituted. Technology no longer uses the verb *to do* without the verb *to care*. To the functionality of the technological artefact is added the affectivity inherent to the possible human purposes it allows to put into practice. It is precisely up to marketing to discover the best advertising strategies to combine the functional and the affective. As Aimee van Wynsberghe points out (2022, 484–485), reciprocity among humans should not be replaced by a faked mutual relationship between humans and robots. The instrumental value of deception, which only seeks to legitimize the acceptance of robotic technology by its users, is, on the contrary, harmful to "the value of reciprocity across society". She argues that blindly accepting an instrumental view of care practices will weaken the capacities of care workers and the political measures that guarantee their professional training (Wynsberghe 2022, 484).

Technological advancements have led to the reinforcement of the problems-solutions discourse. Every new artefact emerges as the prompt solution to a specific problem or, in other cases, for it never to exist. The same is true for other areas of society as well. Every new government measure is conceived and interpreted as a solution to a given problem by the political system. However, we can't have a fruitful awareness of the inherent vulnerability of beings and their environment if we limit our worldview to a technocentric one. No care, no otherness. Care is necessary to preserve otherness.

References

Braga, Joaquim. 2021. Time, space, and care. In *Philosophy of care. New approaches to vulnerability, otherness and therapy. Advancing global bioethics*, ed. Joaquim Braga and Mário Santiago de Carvalho, vol. 16, 129–140. Cham: Springer.

Coeckelbergh, Mark. 2012. Care robots, virtual virtue, and the best possible life. In *The good life in a technological age*, ed. Philip Brey, Adam Briggle, and Edward Spence, 281–292. New York: Routledge.

Gadamer, Hans-Georg. 1993. *Uber die Verborgenheit der Gesundheit: Aufsatze und Vortrage*. Frankfurt am Main: Suhrkamp.

Gadow, Sally A. 1988. Covenant without cure: Letting go and holding on in chronic illness. In *The ethics of care and the ethics of cure: Synthesis in chronicity*, ed. J. Watson and M.A. Ray, 5–14. New York: National League for Nursing.

Heidegger, Martin. 1987. *Zollikoner Seminare: Protokolle, Gesprache, Briefe*. Frankfurt: Klostermann.

Howell, Joel D. 1999. The physician's role in a world of technology. *Academic Medicine* 74 (3): 244–247. https://doi.org/10.1097/00001888-199903000-00013.

Ihde, Don. 1990. *Technology and the lifeworld: From garden to earth*. Bloomington: Indiana University Press.

Kildare, Cory A., and Wendy Middlemiss. 2017. Impact of parents mobile device use on parent-child interaction: A literature review. *Computers in Human Behavior* 75: 579–593. https://doi.org/10.1016/j.chb.2017.06.003.

Meacham, Darian, and Matthew Studley. 2017. Could a robot care? It's all in the movement. In *Robot ethics 2.0: From autonomous cars to artificial intelligence*, ed. Patrick Lin, Keith Abney, and Ryan Jenkins, 97–112. New York: Oxford University Press.

Olesen, Finn. 2006. Technological mediation and embodied health-care practices. In *Postphenomenology: A critical companion to Ihde*, ed. Evan Selinger, 231–245. New York: State University of New York Press.

Radesky, J.S., C.J. Kistin, B. Zuckerman, K. Nitzberg, J. Gross, M. Kaplan-Sanoff, et al. 2014. Patterns of mobile device use by caregivers and children during meals in fast food restaurants. *Pediatrics* 133 (4): 843–850. https://doi.org/10.1542/peds.2013-3703.
Sen, Robin. 2016. Not all that is solid melts into air? Care-experienced young people, friendship and relationships in the 'digital age'. *The British Journal of Social Work* 46 (4): 1059–1075. https://doi.org/10.1093/bjsw/bcu152.
van Wynsberghe, Aimee. 2022. Social robots and the risks to reciprocity. *AI & SOCIETY* 37: 479–485. https://doi.org/10.1007/s00146-021-01207-y.

Chapter 13
Geoengineering as Techno-care: Ethical Issues in/for the Anthropocene

Agostino Cera

Abstract This paper deals with the ecological interpretation of "care" within the Anthropocene framework, which I have called *Techno-care* My basic assumption is that the Anthropocene equates to more than an aspirant geological epoch: it is the best candidate for becoming the *métarécit* of our age. In its essence, the Anthropocene does not correspond to the "Age of Man", but to the epoch in which *technology becomes the* "*subject of history and of nature as well*." Hence the idea that its appropriate name is not *Anthropo*-cene, but *Techno*-cene. Concretely, the following pages sketch the *Paradox of Omni-responsibility*, an ethical paradox emerging within the Anthropocene framework and especially evident in the geoengineering solutions being proposed for ecological problems. I have characterized this oxymoron as *techno-care,* that is, something that transforms responsibility (and care) into a risk rather than a resource. On this basis, the most significant outcome of the Paradox of Omni-responsibility as techno-care is the undermining of Hans Jonas' principle/imperative responsibility as a standard for the ecological thought of recent decades. In the conclusion, I suggest the possibility of a *Care Imperative* as a reply to this undermining.

Keywords Techno-care · Anthropocene · Responsibility · Philosophy of technology · Geoengineering · Pet-ification (of nature)

A. Cera (✉)
Università di Ferrara, Ferrara, Italy
e-mail: agostino.cera@unife.it

J. Braga (ed.), *Ethics of Care*, Advancing Global Bioethics 20,
https://doi.org/10.1007/978-3-031-73608-7_13

Premise

This paper deals with the ecological interpretation of "care" within the Anthropocene framework,[1] which I have called *Techno-care*.[2] According to this interpretation, care represents the benchmark of the relation between human being and world/nature, the barycentre of the link between *anthropos* and *oikos*.

The basic assumption of this work is that the Anthropocene equates to more than an aspirant geological epoch: it embodies the best candidate for becoming the *métarécit* of our age, the most faithful expression of the current *Zeitgeist*. In its essence, the Anthropocene does not correspond to the "Age of Man", "Human Era" or "*Menschenzeit*", but to the epoch in which *technology becomes the "subject of history"* (Anders 2013: 271–79) *and of nature as well*. Insofar as technology becomes both *Weltanschauung* (subject of history) and *Weltform* (subject of nature), it achieves the status of *integral epochal phenomenon*, producing the advent of a *technocosm* where technology represents the *neo-environment*—new *oikos* or vital space—for human being.[3] According to this techno-centred hermeneutical perspective, the Anthropocene counts as the entelechy of the *technisches Zeitalter* (age of technology) outlined by the classical philosophy of technology or the "first generation of philosophers of technology" (Achterhuis 2001: 3). Hence my conviction that the appropriate name for this epochal planetary framework is not *Anthropo*-cene, but rather *Techno*-cene.

With reference to the conference which inspired these pages—the *second IEF International Colloquium "Roads to Care"*, which took place October 19–21, 2022 at the FLUC (Faculty of Arts and Humanities) of the University of Coimbra[4]—I am going to tell the story of a *dead-end for care*, a *short-circuit of the caring society*. Having said that, my intention is not to diminish the meaning of this "new arrival in the history of philosophy" (Pierron 2021), or the current *care turn* in philosophy. More precisely, I do not seek to debase the emergence of a care paradigm within moral philosophy which consists in the idea that, to quote Virginia Held, "the ethics of care is a distinct moral theory or approach to moral theorizing, not a concern that can be added on to or included within other more established approaches."[5] The only aim of these pages is to warn against the risk of an indiscriminate (uncritical)

---

[1] This paper is part of a wider work on the Anthropocene I have been carrying out for some years and that recently became a book (see Cera 2023). This book represents therefore the natural *pendant* of the following pages. On ecological/environmental care see (Curtin 1991); (King 1991); (Postma 2006); (Puig de la Bellacasa 2017:169–215).

[2] In these pages I will agree with the standard definition of care provided by Joan Tronto and Bernice Fischer, according to which care is "a species of activity that includes everything we do to maintain, contain, and repair our 'world' so that we can live in it as well as possible. That world includes our bodies, ourselves, and our environment" (Fischer and Tronto 1990: 40).

[3] With reference to technology as a *neo-environment* and, more generally, the proposal of a *Philosophy of Technology in the Nominative Case* (TECNOM) see (Cera 2017).

[4] See https://www.uc.pt/fluc/uidief/act/roads-to-care/ (accessed October 10, 2024).

[5] Held 2006: 3. About the care turn in philosophy see the already classic (Bowden 1997); (Bubeck 1995); (Held 1995) (and 2006); (Kittay 2002); (Mayeroff 1971); (Noddings 1984) (and 2002); (Pulcini 2009); (Tronto 1993) (and 2013) (and 2015). And more recently: (Braga and Carvalho

openness to care as if it were a kind of moral panacea. My point is that scepticism is and remains the right dress for philosophy: not only its natural *habitus*, but its necessary/essential *ethos*.

Concretely, I will sketch the *ethical paradox (the Paradox of Omni-responsibility) emerging within the Anthropocene framework* that is especially evident in the so-called geoengineering solutions to ecological problems. I have called this oxymoron, which transforms responsibility (and care) from a resource into a risk, techno-care. On this basis, the most significant outcome of the Paradox of Omni-responsibility as techno-care can be found in the overcoming of Hans Jonas' principle/imperative responsibility as a standard for the ecological thought of recent decades. In doing so, I take responsibility not as a synonym of care—i.e. I don't support an equivalence between responsibility and care—but as a fundamental part of the care *dispositif* as established by Berenice Fisher and Joan Tronto, who consider responsibility a constitutive element of care, especially in the phase of "caring for".[6]

My argument consists of three stages. The first deals with a *critical dissection of the Anthropocene as a "discourse"*; the second presents the three main outcomes of the anthropocenic discourse or the *articulation of the techno-care dispositive*: Petification of Nature (i.e. an unprecedented version of its reification), Aidosean Prometheanism (i.e. a new form of anthropocentrism) and the already mentioned Paradox of Omni-responsibility. In the third and final stage, I will draw my conclusions by suggesting the possibility for a *Care Imperative*.

1 From an (Geological) Epoch to a (Philosophical) Discourse

As is well-known, the term "Anthropocene" indicates an aspirant new geological epoch, the third (after Pleistocene and Holocene) of the Quaternary period. According to Paul Jozef Crutzen, the father of this hypothesis, the Anthropocene establishes that "human activity is largely responsible for this exit from the Holocene, that is, *humankind has become a global geological force* in its own right." (Steffen et al. 2011b: 843). In other words, "Human activities have become so pervasive and profound that they rival the great forces of Nature and are pushing the Earth into planetary *terra incognita* [unknown land]" (Steffen et al. 2007: 614).

2021); (Brugère 2021); (Collins 2015); (Groys 2022); (Hamington and Flower 2021); (Puig de la Bellacasa 2017); (The Care Collective 2020); (Urban and Ward 2020).

[6] As is well known, Fischer and Tronto identify four phases in the process of care, namely "caring about", "caring for", "caregiving" and "care receiving" (see Fisher and Tronto 1990). In particular, the "caring for" phase is defined as "the phase in caring when someone assumes responsibility to meet a need that has been identified. Simply a need for care is not enough to make care happen; someone has to assume the responsibility for organizing, marshaling resources or personnel, and paying for the care work that will meet the identified needs. The moral dimension of caring for is to assume, and to take seriously, responsibility" (Tronto 1998: 16–7).

In addition to helping us imagine the concrete future of our planet, this vivid image of *terra incognita* is also valid from a conceptual point of view. In fact, well beyond the intentions of its most enthusiastic supporters (the so-called "Anthropocenologists"),[7] the Anthropocene was quickly revealed to be not so much an epoch as a "discourse," the latter understood in the strong sense of "organizing the perception of a world picture (past, present, and future) through a set of ideas and prescriptions" (Crist 2016: 24). As argued by the historians of science Christophe Bonneuil and Jean-Baptiste Fressoz, l'*Evénement Anthropocène*—i.e. the Anthropocene as a "geohistorical event"—establishes a new *grand récit* where the human being confirms its whole power within "a hegemonic system for representing the world as a totality to be governed" (Bonneuil and Fressoz 2017: 48).

As said, the first stage of my argument consists precisely in a *critical dissection of the Anthropocene as a "discourse"*. My point is that the Anthropocene emerges as an ambiguous idea or an epistemically unstable structure, for it shows an intrinsic tendency to overflow its original epistemic boundaries: those of the natural sciences (geology). As the political scholar Jeremy Baskin argues, the Anthropocene equates less to a scientific concept than to "the ideational underpinning for a particular worldview", even an ideology or "a paradigm dressed as epoch" (Baskin 2015: 9). According to my techno-centred hermeneutics, the ideological character of the Anthropocene depends on the fact that it uncritically accepts the complete *naturalization of technology* as an epoch-making matter of fact. The normative element of such a discourse lies in its unquestioning acceptance of the metamorphosis of *techne* in *physis*. In other words: here and now technology has taken on such a pervasive role that it can only be perceived and interpreted as nature itself. As being *physis*. However, looking at this metamorphosis more closely, it turns out to be the effect of another cause. *Techne* can be understood as *physis* only because it has previously and surreptitiously replaced *physis* in both meaning and function. The metamorphosis of *techne* in *physis* (the naturalization of technology) emerges therefore as an *epiphenomenon* in relation to the main *phenomenon* consisting in the *preliminary* metamorphosis of *physis* in *techne* (the *technologization of nature*), namely in that long process of de-physization/de-cosmization of nature—I refer here to Max Weber's "disenchantment of the world" or Karl Löwith's "deworldification of the world"[8]—which characterizes all of modernity. I call this phenomenon the anthropocenic *Urphänomen*, in the Goethean sense of this word.[9]

On the basis of this assumption, the Anthropocene can be seen as the *redde rationem* of modernity, the moment in which the de-cosmization of nature/disenchantment of the world finds its fulfilment. Within the Anthropocene framework,

[7] The term "Anthropocenologists" refers to the "phalanx of renowned scholars who made the bold gesture of naming our epoch" (Bonneuil and Fressoz 2017: 49).

[8] See, respectively (Weber 2008: 35) and (Löwith 1986: 10).

[9] Goethe calls some phenomena "*Urphänomen*" (archetypal phenomena) because "nothing higher manifests itself in the world; such phenomena, on the other hand, make it possible for us to descend, just as we ascended, by going step by step from the archetypal phenomena to the most mundane occurrence in our daily experience" (Goethe 1983: 195—§ 175).

techne acquires the status of *kosmos* and *holon*; technology becomes an integral epochal phenomenon and thus a totality. What is taking place here, as a result, is not an encounter between *physis* and *techne*, but rather a (con)fusion, or better, a power relation between them. That is to say, an appropriation of nature by technology. This means that the only view yet available to us is that of a "Technature" (see Schwägerl 2014: 127–49), in the sense of "a physics without *physis* and a nature without *logos*" (Löwith 1986: 62). As Crutzen himself affirms in a sort of slip of the tongue, with the coming of the new epoch "it's no longer us against nature", but only because "*nature is* [now] *us*" (Crutzen and Schwägerl 2011—my italics).

2 From a (Philosophical) Discourse to a (Ethical) Dispositif: The Techno-care

At this point, using this first part of my argument as a premise, I would like to emphasize what I consider to be the most "uncanny"—in the Freudian sense of the word—outcome of the Anthropocene from a philosophical and, to be more precise, ethical point of view. It is the unique peculiarity of *anthropocenic care*, *what* I have called *techno-care*.

2.1 The Pet-ification of Nature

It is a matter of fact that one of the main difficulties related to our understanding of this unprecedented epochal framework lies in keeping all its different and seemingly conflicting aspects together, such as the exponential growth of technology, Promethean drive, economic development, social justice, ecological anxieties and so on. All these aspects are summarized in the all-inclusive concept of "sustainability".

My point is that the "unsustainable lightness of sustainability" produces a radical metamorphosis in the images of human being and nature, as well as in their relation. Within the Anthropocene/Technocene scenario, the human being leaves its traditional role as the lord (subject) of a nature conceived as servant (i.e. object or "standing-reserve", to refer to Heidegger's idea of *Bestand*) and takes on that of the "Steward of the Earth System" or "Planetary Manager" (Steffen et al. 2011a) of a nature conceived as living being (see Lovelock and Margulis' Gaia hypothesis, for instance). More precisely, nature takes the form of a *pet*: a living thing, but entirely dependent on our capability to take care of it. Nature is thus something for which we must feel totally, absolutely responsible. As the father of the imperative of responsibility Hans Jonas, argues, "It is his [man's] excess of power that confers this duty on him, and it is against this very power [. . .] that his protection is asked. Thus, it comes about that technology, this coldly pragmatic work of human cunning, installs man in a role which only religion has sometimes assigned to him: that of *steward or guardian of creation*" (Jonas 1982: 894–95—my italics).

Yet here lies the peculiarity of anthropocenic care. The Anthropocene gives birth to a change in nature's image (in our representation of it), which from a warehouse of exploitable resources—a *topos* of modernity—becomes a living entity we must take care of. As a consequence, the basic relationship between human being and nature is no longer characterized by power and domination but by respect and responsibility, that is care (taking). I call this phenomenon the *Pet-ification of Nature*.

Pet-ification of Nature

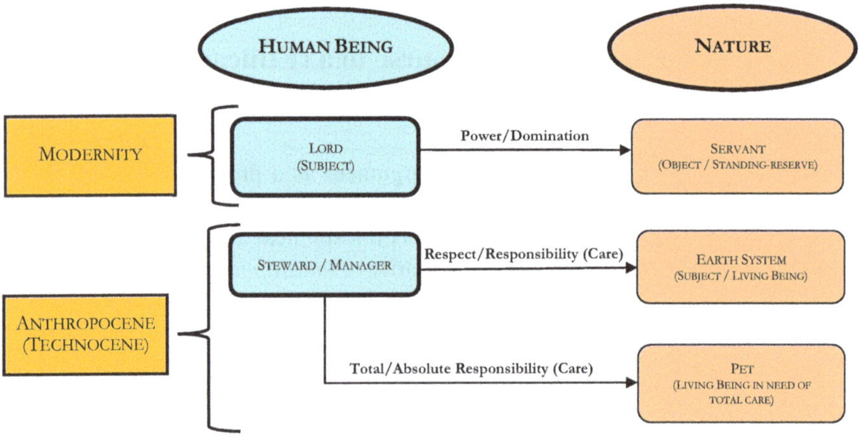

Looking a little deeper, however, nature's new configuration as a pet—and our new role of planetary steward/manager/guardian—presents an intrinsically contradictory character. Beyond its comforting appearance, the evolution of nature's image from inert entity/object to living being/subject, does not automatically mean the overcoming of its reification. This is why I consider the pet-ification of Nature a *trans-objectual reification* of nature. Masked by the nobility of its intentions, such an evolution ends up becoming the perfect strategy for reinforcing nature's reification, the ideal alibi for realizing the definitive annihilation of its otherness. Seen from this perspective, the pet-ification of nature looks like the fulfilment of a more general process in which nature moves from object to Gaia and Earth System.[10] Despite their significant differences, in fact, these interpretations all express the disenchantment of the world as their basic principle; all of them confirm the gradual de-worldification of the world or de-cosmization of nature, namely the systematic erosion of its otherness until its annihilation. These interpretations are, as a result,

[10] According to Paul Crutzen and Will Steffen, the Earth System is "the suite of interacting physical, chemical and biological global-scale cycles and energy fluxes that provide the life-support system for life at the surface of the planet [. . .] the Earth System includes humans, our societies, and our activities; thus, humans are not an outside force perturbing an otherwise natural system but rather an integral and interacting part of the Earth System itself" (Steffen et al. 2007: 615).

all woven into the very fabric of modernity; each forms a chapter in the modern *métarécit*. The leitmotiv of this story is the aforementioned anthropocenic *Urphänomen*, that is nature becoming a function of culture, world a function of human world. According to this approach, pet-ification embodies the final act of the modern *métarécit*.

2.2 The Ways of Reification

In concrete terms, modernity pursues its goal for the disenchantment of the world and the annihilation of nature's otherness by *two ways of reification*: (1) *objectification*, that is, by interpreting the world/nature as an *object*; (2) *pet-ification*, that is, by interpreting the world/nature as a *standing-reserve*.

To avoid a lexical misunderstanding, by "reification" (from the latin *res* = thing) I mean the general instrumentalization of an entity, its generic transformation into a thing, namely a means or a tool for the achievement of an end outside of itself. On the other hand, by "objectification" (from the latin *objectum* = object) I mean a specific kind of reification, one of the two forms this instrumentalization takes. With objectification, the thing becomes object, *res* is transformed into *objectum* or into something "polemically"—in the etymological sense of *polemos* (war, conflict)—opposed to a *subjectum*.

1. *Objectification* (*objectual reification*). The idea of nature/world as an object stands for the final outcome of its theological-Christian interpretation as *ens creatum* and corresponds to the first and most basic form of the disenchantment project. *Objectification is the first modus of (modern) reification*. Here nature's otherness is simply ignored. This approach culminates in the birth of industrial/fossil capitalism, where objectification becomes a worldview. As *objectum*, the world/nature is also *machina* (the Cartesian *res extensa*), an inert dead body/mechanism that represents both a means and an obstacle for *subjectum*'s desiderata. It is therefore something to subjugate and exploit *ad libitum*. Within such a framework there is no place for respect or care; no ecology is possible. Precisely because it is an obstacle, however, the nature-object maintains a trace of autonomy. An object, in fact, is by definition what is "in front of," what puts itself in opposition (see the German "*Gegen*-stand" in addition to the latin "*ob*-jectum"), what the subject must face. Insofar as our relationship with nature is considered to be a struggle, with conquest as its final goal, nature must at least be given the autonomous status of adversary, enemy, counter-force. As a result, in this first case/modus we are faced with an *objectual reification* of nature, that is, a *hard reification* where the denial of nature's otherness occurs in the form of negation by way of a radical contraposition. It is a *struggle* for which a warlike logic (i.e. competitive/ polemic) is adopted.
2. *Pet-ification* (*trans-objectual reification*). After the world/nature as an object, comes the world/nature as a standing-reserve, a well-known Heideggerian term.

> This is the *second modus of (modern)* reification. In his essay *The Question Concerning Technology*, Heidegger establishes the *Gestell* paradigm: the interpretation of technology as an epochal phenomenon, as the "Enframing" within which our current being-in-the-world takes place. Whereas Greek *techne* was characterized by "bringing-forth" (*Hervorbringen*), he uses "standing-reserve" (*Bestand*) to mean the interpretation of entities, which results from the "challenging" (*Herausforderung*) approach typical to modern technology. In his view, "the revealing that rules in modern technology is a challenging (*Herausfordern*), which puts to nature the unreasonable demand to supply energy that can be extracted and stored as such" (Heidegger 1977: 14). Incidentally, these lines would be a perfect epigraph for fossil capitalism. "The name 'standing-reserve,'" so continues Heidegger, "designates nothing less than the way in which everything presences that is wrought upon by the challenging revealing [. . .] Whatever stands by in the sense of standing-reserve no longer stands over against us as object" (Heidegger 1977: 17). Standing-reserve therefore corresponds to a reification of nature's otherness in which the latter is deprived of that last remaining bit of autonomy and difference contained in its interpretation as an object, that is, as a counter-force and obstacle. Now nature really becomes nothing in itself; it is simply a resource, a prosthesis of our desiderata, a mere occasion for exercising our will to power. This is why Heidegger characterizes standing-reserve as "objectless" (*gegenstandlos*), or better, the no-longer-neither-object.

In my opinion, both the Gaia hypothesis and Earth System idea fall under this second modus of (modern) reification. Many would argue, however, that it is precisely these interpretations of the world/nature that try to recognize its otherness. Thanks to these interpretations, they would say, the world/nature is elevated to a subjective status or, at the very least, endowed with a biological dignity. It emerges as something alive and thus automatically as something worthy of consideration, respect, and protection. Within this framework ecology—understood as "care for nature"—can finally find its place. Yet the world/nature pays a steep price for access to the realm of the living. The recognition of nature's otherness in the form of biological dignity depends on its adherence (submission) to, knowingly or not, the logic of techno-science. It is completely reduced to a set of scientific evidences and computable data. The establishment and recognition of Gaia and Earth System are exclusively computational; they are *reified othernesses* and thus oxymorons de facto. Their "vital" parameters and "subjectivity" are entirely convertible into figures, numbers, and data outside of which they cease to exist.[11] This means that in and of themselves—i.e., beyond the legitimacy we give them with our objective knowledge—those othernesses mean nothing, or better, are nothing. The premise underlying these approaches therefore remains the same as reification in general: the denial of the other, the annihilation of its otherness.

---

[11] As Bonneuil and Fressoz rightly argue, "quantifying nature is today big business" (Bonneuil and Fressoz 2017: 55).

In fact, given that what is reduced to calculation and computation is by definition something potentially manageable, the question becomes: "What type of otherness can we be in total command of, reduce entirely to ourselves, namely to our technological-managerial capability?" Clearly it is a fictitious otherness. A simulacrum of otherness. This phantasmatic acknowledgment of nature's otherness proves to be an ideal Trojan horse for its integral domestication and full disenchantment through the pet-ification of nature. We decide to reduce nature's otherness entirely to ourselves in the name of our responsibility, or better, of our *ethical omni-responsibility*, which results from our *technological omni-power*. It is up to us to take care of nature (environment, Gaia, Earth System, biosphere, etc.) because of its intrinsic vulnerability; a vulnerability which is, not surprisingly, inversely proportional to our agency (i.e. our technological omni-power). We thereby establish at the same time both our right to absolute protection (our *total care-giving*) and nature's corresponding, irreversible state of minority (its *total care-needing*, so to speak).

In the name of these noble intentions, the world/nature is placed completely under our protection and thus annihilated as such, denied in its being for itself. Insofar as this integral domestication/full disenchantment occurs in the presence of an entity that is *no longer* conceived as an object (as we've seen, standing-reserve = the no-longer-neither-object), pet-ification represents a *trans*-objectual reification of nature. In fact, differently from objectual reification, which simply ignores nature's otherness, here said otherness is apparently acknowledged, but only on the condition that nature is converted into something completely manageable by the technological agency. We are therefore faced, in this second case/modus, with a *soft reification*: the denial/disenchantment of nature's otherness no longer occurs by way of negation but by erosion and emptying, that is, by *domestication*. It no longer takes the form of a struggle, but rather of *caretaking*, with the warlike logic being replaced by a moralistic, paternalistic one.

The Ways of Reification

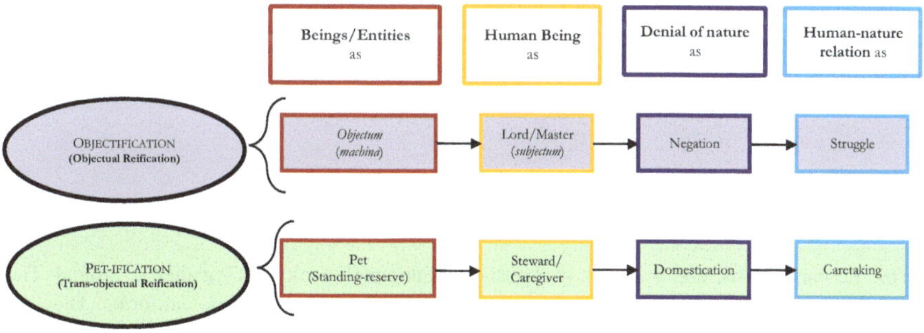

As should be clear by now, the ethical ambiguity emerging from this unprecedented scenario lies in the fact that the stewardship of the Earth System (i.e. our caretaking of the planetary pet) equates to an "active *planetary management*" (Baskin 2015: 10), our absolute right—even duty—to an integral protection of nature. Different ways to address this topic can be found in Anthropocene Studies. The Ecomodernists and other exponents of the Good, Great, or even Beautiful Anthropocene argue that our planetary management should translate into an extensive geoengineering program.[12] Their technolatric motto/dogma, that aims of "decoupling human development from environmental impacts" (Asafou-Adjaye et al. 2015), could sound: "against technology (its excesses), more technology is needed". In doing so, they express a pharmacological view of technology (in the Stieglerean sense of the word)[13] according to which technology represents both the problem and the potential solution, and thus a totality, something that cannot be escaped, namely a destiny. The condition of possibility for establishing a pharmacological view of technology is the existence of a technocosm.

Geoengineering is commonly divided into two "basic 'classes': (1) Carbon Dioxide Removal (CDR) techniques, which remove CO2 from the atmosphere; (2) Solar Radiation Management (SRM) techniques, which reflect a small percentage of the sun's light and heat back into space" (The Royal Society 2009: ix). One of the most incisive and controversial examples of this "techno-fix" (Ellis 2018: 152) is the "artificially adding of aerosols or stratospheric aerosol injection" as a solution to global warming. As is well known, this involves the use of a "bombing" process to artificially increase the concentration of sulphur particles in the stratosphere. The goal of this SRM technique is to cool the planet and balance out the warming caused by greenhouse gas emissions—these too caused by human action (Steffen et al. 2011b: 858).[14] Unfortunately, it seems that geoengineering is going to become the first concrete example of *cosmotechnics*, something very different from the hopeful meaning which Yuk Hui gave to the term.[15]

This kind of strategies proves that he who feels the burden of total responsibility—i.e. the Steward of the Earth System, nature's caregiver—also feels the obligation (the moral imperative) to *fully* exercise it. In other words, within this new epochal context, human being's only potential fault corresponds to *lack of responsibility*, which in turn amounts to abstention. The only guilt still valid in the age of total responsibility is the lack of agency. As a result, geoengineering is not simply a "techno-*fix*", but rather a techno-*care* and thus an oxymoron, something inherently ambiguous from an ethical perspective.

---

[12] The Ecomodernists are a group of "post-environmental ecologists" gathered around The Breakthrough Institute, a think-tank founded in 2007 and based in Oakland, California. They are authors of a very provocative manifesto (Asafou-Adjaye et al. 2015).

[13] Stiegler dealt with care in (Stiegler 2010), in the form of intergenerational care, and (Stiegler 2017), in the framework of philosophy of technology.

[14] According to Erle Ellis, "solar geoengineering using stratospheric sulphates is a shining example of solving one problem by creating an even bigger one" (Ellis 2018: 152). On the ethical implications of geoengineering see (Preston 2016).

[15] On this topic see (Hui 2016) and (Hui and Lemmens 2021).

2.3 The Paradox of Omni-Responsibility (Techno-care's entelechy)

The combination of Pet-ification of nature (i.e. the characterization of planetary organism as a pet) and absolutization of responsibility of the Steward of the Earth System gives rise to the *Ethical Paradox of Omni-Responsibility*, namely the entelechy of the techno-care *dispositif.*

Insofar as the Anthropocene establishes the transition of the human being from "lord over nature" to "Steward of the Earth System", we have the impression that anthropocentrism could lose its traditional negative characterization and gain a new, potentially positive one. This word may no longer signify the blind egoism of a single species at the expense of the framework (the world, the ecosystem), which accommodates it, but that same species' awareness of the authentic implications of its behaviour. As a result, anthropocentrism's natural *pendant* ceases to be *hybris*, namely arrogance, and becomes *ananke*, namely necessity, responsibility and even moral duty. It now corresponds to our duty to take care of and manage the world organism for which we have become entirely responsible, because "The Earth [is] in our Hands" and "it's we who decide what nature is and what it will be" (Crutzen and Schwägerl 2011). Describing his idea of a "new anthropocentrism", Clive Hamilton argues: "the essential lesson of Anthropocene science" is that it is too late "to abandon an anthropocentric standpoint [...] we now have a responsibility for the Earth as a whole and pretending otherwise is itself irresponsible." This means that the question is not "whether human beings stand at the center of the world, but what kind of human being stands at the center of the world, and what is the nature of that world" (Hamilton 2017: 42–3).

While on the surface this may seem encouraging, I think things are quite different. On the basis of its moral and/or ecological duty of total caretaking of its own environment (that is, a situation where technological capability to make something turns into moral obligation to do so), human being gives birth to a *new form of anthrocentrism*: a *neo-Prometheanism*, which proves to be no less problematic than the traditional one. As a consequence, within the Anthropocene framework we are faced with *two kinds of Prometheanism* or anthropocentrism.

1. On the one hand we have the *traditional Prometheanism/anthropocentrism*: the outcome of a *Faustian man*, who considers himself as the lord/subject of a nature conceived as object or standing-reserve. In this case, the *Promethean hybris* is the obvious result of *dis*-interest and *ir*-responsibility (or *un*-care) towards the otherness of nature.
2. On the other hand we have the *anthropocenic or neo- Prometheanism/anthropocentrism*: the outcome of an *Aidosean man* (after Aidos, the Greek goddess of modesty and humility), who considers himself 'only' the steward of a nature conceived as a living being. However, as said, this living being is unable to look after itself, i.e. in permanent need of *care*. In this second case, *Promethean hybris* emerges as the paradoxical result of *hyper*-interest and *omni*-responsibility (or

hyper–/omni- care) towards the otherness of nature. It therefore represents a classic case of heterogony of ends.

The Two Prometheanisms/Anthropocentrisms

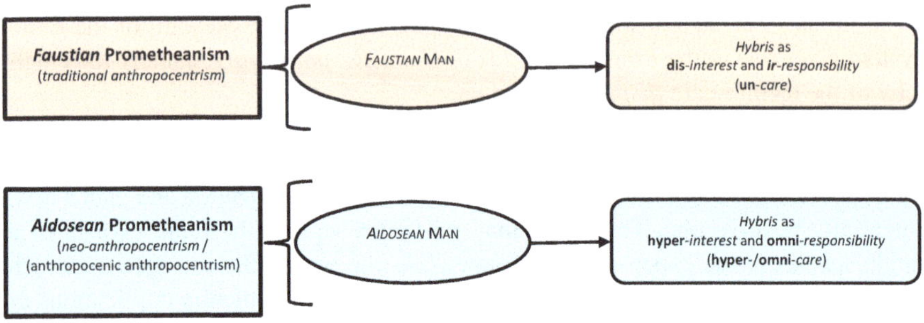

3 Conclusion: Towards a Care Imperative

As we have seen, the main consequence of the Anthropocene understood as *Weltanschauung* and *métarécit*—i.e. as Technocene—is the Pet-ification of Nature, namely the metamorphosis of the idea/image of nature from an object/standing-reserve (*something* to be exploited) into a pet ("*someone*" needing total/absolute care). In turn, the Pet-ification, as a trans-objectual reification of nature, proves to be the entelechy of the tecno-care *dispositif*. This consideration suggests that the Anthropocene undermines Hans Jonas' imperative of responsibility as the ecological (and more in general, ethical) standard of recent decades. If this imperative has been the cornerstone for an "ethics for the technological age" (Jonas 1984: 1), it probably cannot play the same role in the rising anthropocenic/technocenic age. As a conclusion of these pages, I would like to clarify this point.

Concretely, Jonas' imperative of responsibility represents a completion of the "ethic of responsibility" (*Verantwortungsethik*) theorized by Max Weber. According to Weber, as is well-known, the disenchantment of the world and the resultant recognition of a "polytheism of values" imposed on our age the duty to find a concrete and viable alternative to the traditional "ethic of conviction" (*Gesinnungsethik*) symbolized by the Kantian approach (Weber 2008: 359–67). In chapter three of the fourth part of his masterpiece, Jonas tries to establish a "theory of responsibility", that is, an ethical *dispositif* translatable into concrete life situations. An applied responsibility, so to speak. For this purpose, he draws a parallel between private and public responsibility, between the emblematic cases of the parent's responsibility for the infant and that of the statesman for their citizens. They represent two "eminent paradigms" of a *total* responsibility. In particular, he claims that the "*timeless archetype of all responsibility*" (Jonas 1984: 130–35) is that of a parent for their child.

According to Jonas' vocabulary, my idea of "*omni*-responsibility" corresponds to the evolution of this *total* responsibility into an *absolute* responsibility. By "absolute" I refer to its etymological meaning, from the Latin *absolutus*, past participle of the verb *absolvere*: "to set free", "to make separate". In my view this kind of responsibility considers itself *absolved*, by virtue of its noble intentions, from any real consideration/respect for the otherness of its "object". *Omni-responsibility is a responsibility that, precisely due to its moral duty for total caretaking of the other, becomes blind to otherness, indifferent to difference.* This means that, paradoxically, absolute responsibility (i.e. a responsibility concretely/technologically able to realize all its good aspirations) emerges as the anteroom of *property*, namely that kind of relation where the other's otherness is completely denied, where the other is transformed into an object of possession. Looking at the Anthropocene/Technocene scenario, the risk is that, once absolute, our current cosmic/total responsibility could turn into a *totalitarian responsibility*, or better, give rise to a *totalitarianism of responsibility or totalitarianism of care* in the form of an "active planetary management", that is, precisely the Ecomodernist's utopia.

With reference to Jonas' paradigmatic example mentioned earlier, one could argue that an omni-responsible parent—one who is absolute, not only total—equates to the figure known in Italian as "*padre-padrone* (father-master)".[16] This is a parent who, by exercising too much protection (i.e. a complete responsibility, integral/care), literally suffocates the child, inhibiting the development of their autonomous personality. Here the parent ends up enslaving the child, and is transformed from guardian into master. On the political level of public responsibility, the situation is practically identical. This is precisely the type of argument employed by any "father of the nation" or "great helmsman of the people" to justify their totalitarian behaviour. Unfortunately, history give us countless confirmations of this observation. A typical "father of the nation" claims that, for the good of his "children" (the "people"), it is his duty to take on a total/absolute responsibility. In doing so, he presumes to perform a service for the collective body in a gesture of supreme altruism, even sacrifice. His narrative paints the people as eternal babies incapable of taking care of themselves; they cannot, therefore, be left to themselves, namely to their own otherness. Here a *pet-ification of political community* takes place, so to speak. *Polis* turns into a zoo; politics becomes a zoopolitics.

Beyond these eminent paradigms, our everyday experience, in particular that of our closest and strongest relations (i.e. interhuman relations), teaches us that the dark side of responsibility/caretaking lies in taking possession of the other, in transforming it into something for us. From this, a kind of general rule can be derived, according to which the more a relationship is capable of respecting the other's autonomy (or, better, the more it resists the natural impulse to appropriate otherness), the more it can be considered healthy, non-toxic, i.e. "complete." In the end, what we call enchantment (of the world/nature, of a person and so on) is the "simple miracle" of otherness/difference; a miracle which shows that there will always be something/someone irreducible to me. No "you" becomes "I"; when it does, it

[16] This is the title of Gavino Ledda's famous novel (see Ledda 1979) as well as the Taviani brothers' film adaptation of it (winner of the *Palme d'or* at *Festival de Cannes* in 1977).

becomes "mine," and thus ceases to be a "you," transforming a relation into a balance of power and possession.

As the age of our absolute—i.e. total and potentially totalitarian—responsibility towards nature, the Anthropocene/Technocene represents the limit case where Jonas' principle/imperative unknowingly becomes the instrument of a new form of anthropocentrism, precisely because it can be fully actualized/realized. Aidosean Prometheanism is the unexpected child born from the embrace between Prometheus and Gaia. Omni-responsibility—the accomplishment of such an imperative—acts as a Trojan horse for a neo-anthropocentrism that, feeling a duty towards everything (or better, being capable of responding for everything), ends up looking after/caring for everything, that is, measuring everything according to its own claim to a total and absolute responsibility.

Ethically speaking, this means that the most urgent demand of our age is to acknowledge *the limits of responsibility/care*, understood as the potentially dangerous consequences of our best intentions once they become completely achievable, entirely makeable. It's important to remark that within this brand-new epochal framework the technological capability to make something becomes moral obligation to do so, that is, *technology turns into an ethical transcendental*, into the condition of possibility for the ethics itself. This *forma mentis* is perfectly expressed by what Günther Anders and Jacques Ellul considered the categorical imperative of the age of technology. They called it "Gabor's Law", which reads: "what can be made, must be made [...] what must be made is inevitable" (Anders 2013: 17) or, more precisely, "given that we can (technologically) do so, we must (morally) do so". Our technological omni-power gives birth to our ethical omni-responsibility and total care as planetary management.

Thus, in the name of these noble intentions, the world/nature could be placed completely under our protection and therefore annihilated as such, denied in its being for itself. This means that it is no longer *homo faber* at the peak of his Promethean arrogance who brings about this integral domestication (i.e. as subjugation of the nature-object for pure egoism), but rather a conscientious planetary steward/manager who acts for the good of nature in the name of a noble (eco-)altruism. Here the contemptuous lord of nature turns into the docile caregiver of the planetary pet. Yet, paradoxically, it is precisely in this reassuring role that the human being succeeds in achieving its *pan-anthropic project* (the realization of an entirely anthropo-centered world) as never before, not even at the height of its Faustian *hybris*.[17] The paradox, what makes techno-care an oxymoron, lies in the fact that *we could become earthmasters in the form of caregivers*; that is to say not by means of "will to power" but of "*will to care*".

[17] By "pan-anthropism" I mean the totalization of anthropocentrism. If anthropocentrism expresses human being's will not to look outside/beyond itself, pan-anthropism equates to the impossibility to do so, that is, human being's supervening inability to look outside/beyond itself. Pan-anthropism makes anthropocentric self-referentiality (its blindness to otherness/difference) something natural, even necessary. On this topic see (Cera 2023: 37–8 and 72).

At the same time our age demands that we become aware of a brand-new ethical problem: the potential *aporia between the responsibility for the other and the respect of its otherness*, namely that no authentic "*Verantwortung*" (responsibility)—or care—is possible without "*Gelassenheit*" (releasement). If not, responsibility/care becomes nothing but management, that is, a kind version of control and domination. All forms of care (tutelage/preservation)—not just the absolute and total one—should safeguard the other's being other, yielding to a superior form of "ir-responsibility" or "un-care": not that of dis-interest, but that of the superior interest expressed by *Gelassenheit*. Such a superior irresponsibility/uncare knows the value of difference as the authentic epiphany of otherness. The care receiver's autonomy itself should be considered a function of their otherness/difference.[18] There is no way to produce, create, or cause difference; we can only welcome it, recognize it, preserve it, offer it hospitality. All we can do is let it happen, let it be.

Here the "simplicity" of lived experience could help us clarify this apparently abstract issue. The most difficult and painful part of being a parent—what Jonas rightly calls the "timeless archetype of all responsibility"—is learning that the apotheosis of love for one's own child is letting them live their own life, letting them go, letting them be, releasing them. This conscious and painful abstention consists in stifling the natural impulse to constantly protect one's own child, to constantly answer for him/her. From Plato's *Republic* to Rousseau's *Emile*, we know that at the basis of even the most highly developed and most nobly inspired educational projects lies the temptation of control and domination. Every *Bildung* risks building a cage. To avoid doing so, every parent must understand and accept that even one's own blood is another blood—a different blood. Every parent must learn that giving life to someone also and perhaps primarily means, at a certain point, leaving that life to them. Answering for someone means stealing their voice, making them silent. To give life to someone essentially means to give life to someone *else*, to *another* person.

Awareness of this unprecedented ethical issue—the aporia between responsibility for the other and respect for its otherness—could suggest the first stage for its overcoming, the first philosophical step toward building an ethical/moral paradigm good enough for this brand-new epoch. In my view, this first stage corresponds to an *encounter between Verantwortung and Gelassenheit* (between responsibility/care and releasement), namely a renewed ethical dialogue between Jonas and Heidegger. While Releasement represents the fundamental guarantee of Difference/Otherness, the latter embodies the condition of possibility for a genuine Responsibility. Thus, in addition to the imperative of responsibility, an *imperative of releasement* is needed; a *Prinzip Gelassenheit* must accompany the *Prinzip Verantwortung*. This is the additional, and at first glance contradictory, counterweight needed by responsibility (as omni-responsibility) to balance the excesses of technological omni-power, to contrast the new *hybris* of hyper-interest and total caretaking.

---

[18] On the topic of care and respect/autonomy see (Davis 1990) and (Dillon 1992).

The renewed encounter between these two principles could also be called *Imperative of Care* or *Prinzip Sorge* and we can already establish its basic rule in two versions, that is, (1) Every caretaking/responsibility must accept the limit of safeguarding the other's otherness, letting it be (releasing it) in its per se. (2) Authentic responsibility must be able to acknowledge and preserve the value of difference as the irreplaceable epiphany of otherness. What emerges in both versions is that the only reliable measure of our responsibility/care is (must be) our "Taste for Other(ness)."[19]

Techno-care is an oxymoron and aporia because technology inherently obeys the values of rationalization and disenchantment, two parameters which disregard the acknowledgment of otherness and alterity. In order to realize a "total mobilization", that is, to become a planetary force or an integral epochal phenomenon, technology needs to make everything interchangeable, but in order to make everything interchangeable technology needs to abolish every kind of difference. It must become indifferent to difference. That is why a responsibility/care measured by the logic of technology can be nothing but nonsense, a contradiction in terms.

Let me conclude by quoting the last words of Joaquim Braga's text *On Caring Acts and Technological Mediation Processes*: "care is necessary to preserve otherness" (Braga 2024).[20] I would even go further and say "care *depends* on the preservation of otherness", in the sense that such a preservation should be established as the goal of an authentic care, even *its condition of possibility*. Because *without otherness*, i.e. without the safeguard of difference, *no care is possible*.

A Heideggerian Postscript

To wrap up this dialogue with Jonas' Imperative Responsibility I would like to add another piece to the suggested Imperative Care by way of a brief "Heideggerian digression". Given that Heidegger can be considered an *ante litteram philosopher of care*, I will translate the above argument concerning the paradox of omni-responsibility into the lexicon of existential analytics: the theoretical core of *Being and Time*.

The ethical paradox I just highlighted could also be described as a short circuit between what Heidegger calls "caring for" (*Fürsorge*) and "concern with" (*Besorgen*): the two most basic ways for expressing "Care" (*Sorge*) as "primordial totality of the structural whole of Da-sein" (Heidegger 1996: 179),[21] as unity in

[19] Here I wanted to evoke a beautiful film by Agnes Jaoui by paraphrasing its title: *Le Goût des autres* (2000).

[20] See also (Braga 2024: 129–40).

[21] "The formal existential totality of the ontological structural whole of Da-sein must thus be formulated in the following structure: The being of Da-sein means being-ahead-of-oneself-already-in (the world) as being-together-with (interwordly beings encountered). The being fills in the significance of the term care, which is used in a purely ontological and existential way. Any ontically tendency of being, such as worry or carefreeness, is ruled out." (Heidegger 1996: 179–80). Chapter 9 of Boris Groys's *Philosophy of Care* deals with Heidegger's *Sorge*, actually a somewhat didactic presentation of the topic (Groys 2022). More interesting are (Paley 2000); (Scott 2010); (Brencio 2022).

which all "existentials" are gathered. "Caring for" ("being-with"), in particular, represents the way in which Da-sein relates to other Da-seins (*Mit-Daseins*), that is, to its "peers," while "concern with" ("being-together-with") denotes how Da-sein relates to things. The extremely peculiar relation established between human being and nature—between *anthropos* and *oikos*—within the Anthropocene/Technocene shows how a totalized/absolutized *Fürsorge* (the kind of *Fürsorge* expressed by omni-responsibility and total caretaking) paradoxically culminates in a *Besorgen*. Total "caring for" turns into a "concern with," that is, into a "having to do with" in the sense of "having at one's disposal." As a result, its ontic correlate—i.e. the referent/counterpart of such a way of doing—is no longer "*Mit-Dasein*" (a peer Da-sein, someone who is "like me" precisely by virtue of their irreducible "being for itself," of their otherness), but rather a "*Vorhandenes/Zuhandenes*": something "present-at-hand," namely a "ready-to-hand" at my total disposal. In other words, this "totally cared for entity" becomes an object or even a "standing-reserve" (*Bestand*). As is well-known, according to *Being and Time*'s vocabulary, *Vorhandenheit* ("presence-at-hand/occurrentness") and *Zuhandenheit* ("readiness-to-hand/ availableness/ handiness") represent the ways of being of things, those entities different from the Da-sein. As a result, the *Fürsorge* which turns into *Besorgen* equates de facto to a *reification*: that kind of behavior which shapes its referent/counterpart as an object or standing-reserve. Such behavior makes an entity be what it is not, forces it to assume a form (way of being) that is not its own. It therefore emerges as an oppressive, violent behavior veiled by attention and caretaking.

All told, this brief incursion into the theoretical *dispositif* of existential analytics seems to be an additional confirmation of our fear: total/absolute care leads to dominion and possession in the form of management and control. *Total/absolute caregiving leads to mastery*. Once omni-responsibility becomes real (that is, concretely feasible, makeable), the abyss of its ambiguity is opened; this was unimaginable when it was "just" a regulatory ideal from which to draw inspiration. So, the establishment of the Technocene proves that behind *Fürsorge* hides the danger of *Besorgen*; behind *Mit-Dasein* lurks the specter of *Zuhandenheit/Vorhandenheit*.

References

Achterhuis, Hans, ed. 2001. *American philosophy of technology: The empirical turn*. Trans. R. P. Crease. Bloomington/Minneapolis: Indiana University Press.

Anders, Günther. 2013. *Die Antiquiertheit des Menschen 2. Über die Zerstörung des Lebens im Zeitalter der dritten industriellen Revolution*. München: Beck.

Asafou-Adjaye, John et al. 2015. *An Ecomodernist Manifesto*. April 2015 http://www.ecomodernism.org/manifesto-english. Accessed 10 Oct 2024.

Baskin, Jeremy. 2015. Paradigm dressed as epoch: The ideology of the Anthropocene. *Environmental Values* 24 (1): 9–29. https://doi.org/10.3197/096327115X14183182353746.

Bonneuil, Christophe, and Jean-Baptiste Fressoz. 2017. *The shock of the Anthropocene: The Earth, history and us*. Trans. D. Fernbach. London/New York: Verso.

Bowden, Peta. 1997. *Caring*. London/New York: Routledge.

Braga, Joaquim. 2024. On caring acts and technological mediation processes. In *Ethics of care: Theoretical and practical perspectives*, ed. Joaquim Braga. Cham: Springer.
Braga, Joaquim, and Mário Santiago de Carvalho, eds. 2021. *Philosophy of care: New approaches to vulnerability, otherness and therapy*. Cham: Springer.
Brencio, Francesca. 2022. Martin Heidegger e il pensiero della cura. *La società degli individui* 73 (1): 35–44. https://doi.org/10.3280/LAS2022-073004.
Brugère, Fabienne. 2021. *L'étique du «care»*. Paris: PUF.
Bubeck, Diemut. 1995. *Care, gender and justice*. Oxford: Clarendon Press.
Cera, Agostino. 2017. The technocene or technology as (neo)environment. *Techné: Research in Philosophy and Technology* 21 (2/3): 243–281. https://doi.org/10.5840/techne201710472.
———. 2023. *A philosophical journey into the Anthropocene: Discovering Terra Incognita*. Lanham/Boulder/New York/London: Lexington Books.
Collins, Stephanie. 2015. *The core of care ethics*. New York: Palgrave Macmillan.
Crist, Eileen. 2016. On the poverty of our nomenclature. In *Anthropocene or Capitalocene? Nature, history, and the crisis of capitalism*, ed. Jason W. Moore, 14–33. Oakland: PM Press.
Crutzen, Paul J., and Christian Schwägerl. 2011. Living in the Anthropocene: Toward a new global ethos. *Yale Environment 360*, January 24, 2011. https://e360.yale.edu/features/living_in_the_anthropocene_toward_a_new_global_ethos. Accessed 10 Oct 2024.
Curtin, Deane. 1991. Towards an ecological ethic of care. *Hypatia: A Journal of Feminist Philosophy* 6 (1): 60–74.
Davis, Anne J. 1990. Are there limits to caring? Conflict between autonomy and beneficence. In *Ethical and moral dimensions of care*, ed. M.M. Leininger, 25–32. Detroit: Wayne State University Press.
Dillon, Robin E. 1992. Care and respect. In *Explorations in feminist ethics. Theory and practice*, ed. E. Browning Cole and S. Coultrap-McQuin, 69–81. Bloomington/Indianapolis: Indiana University Press.
Ellis, Erle C. 2018. *Anthropocene: A very short introduction*. Oxford: Oxford University Press.
Fisher, Berenice, and Joan Tronto. 1990. Toward a feminist theory of caring. In *Circles of care*, ed. E. Abel and M. Nelson, 35–62. Albany: SUNY Press.
Goethe, Johann Wolfgang von. 1983. Theory of Color. In *Scientific Studies (Goehte Collected Works vol. 12)*. Trans. D. E. Miller, 157–298. New York: Suhrkamp.
Groys, Boris. 2022. *Philosophy of care*. London/New York: Verso.
Hamilton, Clive. 2017. *Defiant earth: The fate of humans in the Anthropocene*. Cambridge/Malden: Polity Press.
Hamington, Maurice, and Michael Flower, eds. 2021. *Care ethics in the age of precarity*. Minneapolis/London: University of Minnesota Press.
Heidegger, Martin. 1977. The question concerning technology. In *The question concerning technology and other essays*. Trans. W. Lovitt, 3–35. New York/London: Garland Publishing.
———. 1996. *Being and time: A translation of Sein und Zeit*. Trans. J. Stambaugh. New York: State University of New York Press.
Held, Virginia, ed. 1995. *Justice and care: Essential readings in feminist ethics*. Boulder: Westview Press.
———. 2006. *The ethics of care*. New York: Oxford University Press.
Hui, Yuk. 2016. *The question concerning Technology in China: An essay in Cosmotechnics*. London: Urbanomic.
Hui, Yuk, and Pieter Lemmens, eds. 2021. *Cosmotechnics: For a renewed concept of technology in the Anthropocene*. London/New York: Routledge.
Jonas, Hans. 1982. Technology as a subject for ethics. *Social Research* 49 (4): 891–898.
———. 1984. *The imperative of responsibility: In search of an ethics for the technological age*. Trans. H. Jonas and D. Herr. Chicago/London: The University of Chicago Press.
King, Roger J.H. 1991. Caring about nature: Feminist ethics and the environment. *Hypatia: A Journal of Feminist Philosophy* 6 (1): 75–89.
Kittay, Eva Feder, ed. 2002. *The subject of care*. Lanham: Rowman & Littlefield.

Ledda, Gavino. 1979. *Padre padrone: My father my master*. Trans. G. Salmanazar. New York: Urizen Books.

Löwith, Karl. 1986. Gott, Mensch und Welt in der Metaphysik von Descartes bis zu Nietzsche. In *Sämtliche Schriften*, Band 9, hrsg. von Henning Ritter, 1–194. Stuttgart: Metzler.

Mayeroff, Milton. 1971. *On caring*. New York: Harper & Row.

Noddings, Nel. 1984. *Caring: A feminine approach to ethics and moral education*. Berkeley: University of California Press.

———. 2002. *Starting at home: Caring and social policy*. Berkeley/Los Angeles/London: University of California Press.

Paley, John. 2000. Heidegger and the ethics of care. *Nursing Philosophy* 1 (1): 64–75. https://doi.org/10.1046/j.1466-769x.2000.00004.x.

Pierron, Jean-Philippe. 2021. Care: A New Arrival in the History of Philosophy? In *Philosophy of Care: New approaches to vulnerability, otherness and therapy*, ed. Joaquim Braga, and Mario Santiago de Carvalho. 3–18. Cham: Springer.

Postma, Dirk Willem. 2006. *Why Care for Nature? In search of an ethical framework for environmental responsibility and education*. Dordrecht: Springer.

Preston, Christopher J., ed. 2016. *Climate justice and geoengineering: Ethics and policy in the atmospheric Anthropocene*. London/New York: Rowman & Littlefield.

Puig de la Bellacasa, María. 2017. *Matters of care: Speculative ethics in more than human worlds*. Minneapolis/London: University of Minnesota Press.

Pulcini, Elena. 2009. *La cura del mondo*. Torino: Bollati Boringhieri.

Schwägerl, Christian. 2014. *The Anthropocene: The human era and how it shapes our planet*. Trans. L. Renner Jones. Santa Fe: Synergetic Press.

Scott, Charles E. 2010. Care and authenticity. In *Martin Heidegger: Key concepts*, ed. B.W. Davis, 57–68. Durham: Acumen.

Steffen, Will, Paul Crutzen, et al. 2007. The Anthropocene: Are humans now overwhelming the great forces of nature? *Ambio* 36 (8): 614–621. https://doi.org/10.1579/0044-7447(2007)36[614:TAAHNO]2.0.CO;2.

———. 2011a. The Anthropocene: From global change to planetary stewardship. *Ambio* 40 (7): 739–761. https://doi.org/10.1007/s13280-011-0185-x.

———. 2011b. The Anthropocene: Conceptual and historical perspectives. *Philosophical Transactions of The Royal Society A Mathematical Physical and Engineering Sciences* 369 (1938): 842–867. https://doi.org/10.1098/rsta.2010.0327.

Stiegler, Bernard. 2010. *Taking care of youth and the generations*. Trans. Barker. Stanford: Stanford University Press.

———. 2017. What is called caring? Beyond the Anthropocene. *Techné: Research in Philosophy and Technology* 21 (2/3): 386–404. https://doi.org/10.5840/techne201712479.

The Care Collective. 2020. *The care manifesto: The politics of interdependence*. London/New York: Verso.

The Royal Society. 2009. *Geoengineering the climate: Science, governance and uncertainty*. London: The Royal Society.

Tronto, Joan. 1993. *Moral boundaries: A political argument for an ethic of care*. New York: Routledge.

Tronto, Joan C. 1998. An ethic of care. *Generations: Journal of the American society on Aging* 22 (3): 15–20.

———. 2013. *Caring democracy: Markets, equality, and justice*. New York/London: New York University Press.

———. 2015. *Who cares? How to reshape a democratic politics*. Ithaca/London: Cornell University Press.

Urban Petr, and Lizzie Ward, eds. 2020. *Care ethics, democratic citizenship and the state*. Cham, Switzerland: Palgrave Macmillan.

Weber, Max. 2008. Science as a vocation. In *Max Weber's complete writings on academic and political vocations*. Trans. G. C. Wells, 25–52. New York: Algora Publishing.

The manufacturer's authorised representative in the EU is Springer Nature Customer Service Centre GmbH, Europaplatz 3, 69115 Heidelberg, Germany. If you have any concerns regarding our products, please contact ProductSafety@springernature.com

Printed and bound by CPI Group (UK) Ltd, Croydon, CR0 4YY
30/09/2026
02225592-0004